CONTEMPORARY SOUTH AFRICAN DEBATES

CONTEMPORARY SOUTH AFRICAN DEBATES

Other titles available in this series:

From apartheid to nation-building
by Hermann Giliomee and Lawrence Schlemmer
ISBN 0 19 570550 5

Critical choices for South Africa: An agenda for the 1990s
edited by Robert Schrire
ISBN 0 19 570574 2

No place to rest: Forced removals and the law in South Africa
edited by Catherine O'Reagan and Christina Murray
ISBN 0 19 570580 7

Towards justice? Crime and state control in South Africa
edited by Desiree Hansson and Dirk van Zyl Smith
ISBN 0 19 570579 3

Protecting human rights in a new South Africa
by Albie Sachs
ISBN 0 19 570609 9

People and violence in South Africa
edited by Brian McKendrick and Wilma Hoffmann
ISBN 0 19 570581 5

The elusive search
for
peace
South Africa, Israel and Northern Island

Edited by
Hermann Giliomee
and
Jannie Gagiano

CONTEMPORARY SOUTH AFRICAN DEBATES

1990
Oxford University Press
Cape Town
in association with
IDASA

Oxford University Press
Walton Street, Oxford OX2 6DP, United Kingdom

Oxford New York Toronto
Delhi Bombay Calcutta Madras Karachi
Petaling Jaya Singapore Hong Kong Tokyo
Nairobi Dar es Salaam Cape Town
Melbourne Auckland

and associated companies in
Berlin Ibadan

ISBN 0 19 570611 0

Published by Oxford University Press Southern Africa
Harrington House, Barrack Street, Cape Town, 8001, South Africa
in association with the
Institute for a Democratic Alternative for South Africa

Set in 10 on 12 pt Garamond by Photoprint
Printed and bound by Clyson Press, Maitland, Cape

Contents

NORTHERN IRELAND

SOUTH AFRICA

TOWARDS ACCOMMODATION

Contributors

Heribert Adam is Professor of Sociology at Simon Fraser University in Vancouver, Canada, and author of *South Africa without Apartheid*.

Neville Alexander was a Humboldt scholar and has been active in extra-parliamentary politics. Shortly after his release from Robben Island he published, under the pseudonym No Sizwe, *One Azania: The National Question in South Africa*.

David Apter is Professor of Comparative Politics and Social Development at Yale University, and author of *Against the State*, a study of political violence in Japan, and of *The Politics of Modernization*.

Meron Benvenisti was formerly Deputy Mayor of Jerusalem, and presently directs an independent research group established to investigate conditions in the West Bank and Gaza. His latest book is *The Shepherds' War*.

John Brewer is Reader in the Department of Social Studies at Queen's University in Belfast. He has written extensively on South Africa, and on policing, especially in South Africa and Northern Ireland.

Walker Connor is Professor of Political Science at Trinity College in Connecticut, United States, and is author of *The National Question in Marxist-Leninist Theory and Strategy*.

Bernard Crick is Emeritus Professor of Birkbeck College, University of London, Honorary Fellow of the University of Edinburgh, and author of *In Defence of Politics* and *George Orwell: A Life*.

Jannie Gagiano lectures at the University of Stellenbosch, and has been reporting on white student attitudes since the early 1970s.

Norman Gibson is Professor of Economics at the University of Ulster, and author of articles on the political economy of Northern Ireland.

Hermann Giliomee is Professor of Political Studies at the University of Cape Town, and co-author of *From Apartheid to Nation-building.*

Adrian Guelke teaches political science at Queen's University in Belfast, and is author of *Northern Ireland: The International Perspective.*

R W Johnson is a fellow of Magdalen College, Oxford, and author of *How Long Will South Africa Survive?*

Michael MacDonald teaches in the Department of Political Science at Williams College in Massachusetts, United States, and is author of *Children of Wrath: Political Violence in Northern Ireland.*

Benyamin Neuberger is Associate Professor in the Department of Middle East and African History at Tel Aviv University, and author of *National Self-determination in Post-colonial Africa.*

Sari Nusseibeh is a scion of a prominent Palestinian family in the West Bank, and lectures in the Department of Philosophy at Bir-Zeit University.

Padraig O'Malley is a Senior Associate at the John W McCormack Institute of Public Affairs at the University of Massachusetts in Boston, and author of *The Uncivilized Wars: Ireland Today.*

Lawrence Schlemmer is Director of the Centre for Policy Studies at the University of the Witwatersrand, and co-author of *From Apartheid to Nation-building.*

Introduction

HERMANN GILIOMEE

There is an abundance of individual studies about South Africa, Northern Ireland and Israel. It is not difficult to see why. In each case two distinctive communities are locked into a conflict full of heroism, tragedy and evil deeds. This drama is playing itself out to a world audience watching with morbid fascination. Both inside and outside the respective countries few remain neutral; the question "whose side are you on?" is never far from the surface. For the Western world the three societies pose a peculiar dilemma: all three are part of the "unfinished business" of European colonisation, and all three claim to be representatives of Western values and as such deserving of support.

Apart from the human drama there are also enormous Western material and security interests at stake. Hence all the analyses assessing the nature of the conflict, the magnitude of discrimination and privilege, and the prospects for peace or a breakdown of the social order. It is particularly the possibility of social breakdown, or what Albert Camus has called the "fatal embrace", that haunts the world. Writing about the final stages of the civil war in his native land, Algeria, Camus sadly observed: "It is as if two insane people, crazed with wrath, had decided to turn into a fatal embrace, the forced marriage from which they cannot free themselves. Forced to live together and incapable of uniting, they decide at last to die together." Subconsciously or not, a good proportion of the books on South Africa, Northern Ireland and Israel are written with the aim of helping to avoid the fatal embrace.

In the public perception there is a vague but nevertheless unshaken belief that a common thread runs through the "troubles" in Ulster, the "unrest" in South Africa and the *"intifada"* in the territories occupied by Israel. However, with notable exceptions, scholars have been curiously reluctant to embark on systematic comparisons which aim at drawing conclusions by viewing South Africa, Northern Ireland and Israel together.

Comparative studies, whether of an institution like slavery or segregation, or of two or more societies, differ in their basic approaches.[1] There are those which stress parallels and which use this context to test certain hypotheses or theories. The work of Stanley Greenberg, *Race and State in Capitalist Development: Comparative Perspectives*, falls in this category.[2] Comparing South Africa, Northern Ireland, Israel and the state of Alabama, Greenberg shows that contrary to modernisation theory and the so-called "growth hypothesis", ethnic divisions and discrimination in those capitalist societies have persisted despite long histories of economic growth and development.

The second kind of comparative analysis is concerned with establishing and explaining contrasts. Analysts in this mould want to find out why two or more societies which have, say, a common history of European colonisation or share a common religion, develop differently. There is a tendency here to argue that the founding impulse and earliest history launch each specific society on a different trajectory which strongly affects present-day actions and responses.

From this perspective, the present conflict in Northern Ireland is still largely shaped by the fact that Ulster was founded as a Protestant plantation loyal to the British crown. The crisis in South Africa, by contrast, is shaped by its history as a colonial society built by private enterprise exploiting native labour. Israel, in turn, faces its Palestinian adversaries as a state built on the Zionist idea and Jewish self-sufficiency in labour.

Against this background and using what has been called the contrast-oriented comparative approach, Heribert Adam argues that the two "settler societies" of Israel and South Africa employ quite different methods to deal with challenges from their subordinate populations. Because of white-black economic interdependence, South Africa has to co-opt ever increasing numbers of blacks into the government and state apparatus. Israel, on the other hand, depends on the support of the Jewish diaspora but not on Palestinian labour. Consequently Israel's favoured solution remains the exclusion of the Palestinians in order to remain a democratic Jewish state.[3]

There is also a third comparative approach. Recognising both similarities and contrasts, scholars embark on a macrosocial inquiry to establish the causes of the different responses of societies to common problems and challenges. The focus is on the commonality of the problem rather than the similarities between or among societies which may (or may not) exist.

Northern Ireland, South Africa and Israel face the same problem: European settler societies which try on the one hand to maintain domination in state and society and, on the other, to retain a measure of democratic respectability in Western eyes. Pierre van Berghe has provocatively captured the phenomenon with the term "herrenvolk democracies",[4] and has used this as a leading theme in a comparison of Israel and South Africa.

In an article published in 1980 Sammy Smooha asked a different question: given the enormous strains inherent in these conflict-form societies, why did Israel up to that point display far greater coherence than Northern Ireland, which was engulfed by civil strife? His answer was that "Israel has managed to keep internal peace owing to its more effective machinery of control and not because the potentiality for conflict and the intensity of conflict between Arabs and Jews are less than those between Catholics and Protestants."[5] Since Smooha's article was published Israel has also experienced a breakdown of cohesion and control similar to what occurred in Northern Ireland after 1968 and in South Africa after 1976.

Relying mostly on the third type of comparative analysis, this book poses the problem: why are the conflicts in Northern Ireland, South Africa and Israel so intractable and why is the search for peace so elusive?

The book originated as a conference held in Bonn, West Germany in September 1989. The conference was funded by the Friedrich Naumann Foundation and was hosted by this foundation in association with the Institute for a Democratic Alternative for South Africa (Idasa). Forty scholars, journalists and activists from nine different countries attended. With the exception of the contribution by Sari Nusseibeh, who could not attend, all the chapters in this book were originally presented at the conference in September 1989 and then revised in the six months following the conference. The final chapter was written in mid-1990.

This book reflects a sober assessment of the situation in the three societies. On the one hand the easy or first-choice solutions like partition, power-sharing or simple majoritarianism have failed. On the other hand mere coercion is no longer effective. Whereas in the past resistance was quickly smashed or disintegrated in the three societies (one thinks of the ignominious failure of the IRA in the late 1950s and early 1960s), a new pattern of conflict seems to have emerged. One now witnesses in all three societies low-key violence leading neither to full system break-down nor to renewed full control. The search for peace remains elusive.[6]

To find new and more promising avenues which could lead to peace this book aims at a fundamental reassessment of the conflict in Northern Ireland, South Africa and the territories occupied by Israel. It looks firstly at the ideological and institutional forces which have shaped the conflict and the obstacles to its resolution before turning to the prospects for accommodation and what Bernard Crick has called the high price which all parties, individually and collectively, have to pay if they really want peace. The great majority of the contributors to this book favour a peaceful resolution of the conflict followed by the introduction of democratic practices as far as these are possible. However, contributors were asked to analyse the problem, not offer a solution.

In the first section of the book, the forces which shape the conflict are analysed by MacDonald, Neuberger, Guelke, Brewer and Connor. MacDonald argues that the conflict would not have been as intractable if it was merely a matter of settlers building up a positive identity as people who loved the land, obeyed the precepts of their religion, maintained their traditions and developed an indigenous culture. As MacDonald observes, the problem is that settlers and natives defined themselves relationally.

Settlers were not only what they were but also what they were not: they were *not* primitives, heathens, heretics or natives, but a special kind of people, perhaps even a Chosen People. Their superiority, their uniqueness depended on the natives' inferiority. To grant the subordinate community equal status was to destroy the distinctiveness of the settler community. Indeed it put at risk its very political survival. The quest became political sovereignty based on the exclusion of the other.

Neuberger examines how subordinate groups develop alternative notions of nationhood, justice and order which stand fundamentally opposed to those of dominant groups. While initially the subordinates pose the struggle in civil rights terms, the dynamic of the conflict turns it into a national liberation struggle in which the oppressed demand to control the state (or get one of their own) and be governed by their own kith and kin.

The most important agencies mediating the conflict are the police force and the international community, the themes of chapters by Brewer and Guelke. As Brewer points out, the police in the societies under discussion are hardly disinterested social arbiters. Indeed, the main feature of policing in divided societies is selective enforcement of the law. Illegal political activity by members of the dominant group is condoned; by contrast, the political behaviour of subordinates is closely monitored, and the letter of the law is used to harrass activists. In a sense it is incorrect to use the term "intercommunal civil war" for the conflicts in Ulster, South Africa or Israel. For it is not communities fighting each other, but the security forces pitted against the civilians of the subordinate community.

Guelke's chapter focuses on the role of the international community. Outbursts of violence galvanise the international community, for they bring the basic political illegitimacy of a regime (or an occupation of an area) into sharp focus. The demand of the international community is for self-determination of the people in a given territory. This means majority self-determination and *not* ethno-national self-determination (as the whites in South Africa or the Protestants in Ulster demand). The international community tends to anoint a resistance movement as the legitimate successor government well before victory is achieved. It is prepared to consider the question of safeguards for erstwhile ruling groups, but only after they have conceded the principle of majority rule.

Finally, Connor's chapter deals with ethnic conflicts from a universal perspective, showing why it is so difficult to resolve them peacefully. Claiming that the conflict is not only, or even in the first place, over the dominant group's preferential access to capital, land, jobs, education and residential areas, Connor warns that scholars greatly exaggerate the influence of the materialistic factor in ethnic conflicts. Equally or, in Connor's perception, more important is the conflict over national identity, or to put it simply: which ethnic or communal group governs, defines the nation and chooses its symbols?

While ruling groups which are in retreat are often prepared to give up domination, they strongly resist, even to the point of armed resistance, becoming subordinate to those over whom they once lorded it. Usually they insist on some form of autonomy but this is often rejected by their antagonists as an offensive reminder of the old order.

The subsequent sections of the book present case studies of Israel, Northern Ireland and South Africa, highlighting the key areas of conflict in the interests and ideologies of the main contending groups.

The section on Israel deals only with the occupied territories of the West Bank and Gaza. At the Bonn conference Sammy Smooha presented his view, also published elsewhere,[7] that the occupation is not irreversible since Israel strongly wants to remain Jewish and democratic. The question is simply whether a viable option will become available behind which the Jewish public can be mobilised.

Benevisti argues that, on the balance of probabilities, there is no ideological force which can reverse the occupation. From his point of view, it is better to prepare for resolving a prolonged intercommunal struggle or even civil war.

Nusseibeh, a Palestinian academic who lives on the West Bank, predicts that if Israel remains intransigent Palestinians may begin to call for integration and embark on a civil rights struggle.

In the section on Northern Ireland, Norman Gibson asks whether a reduction in the privileges and power of the dominant group makes the conflict less acute. The chapter shows that despite the progress made over the past two decades in removing discrimination against Catholics, the exclusiveness of the groups and their antagonism towards each other remain as stark as ever. Gibson, who comes from Northern Ireland and a Protestant background, points out that when the survival of a distinctive culture and identity becomes an end in itself the conflict becomes a zero-sum game in which negotiation and compromise are seemingly impossible.

Padraig O'Malley, born and raised in a Catholic community in the Republic of Ireland, analyses the Anglo-Irish Agreement of 1985 which the British and Irish governments hoped would break the deadlock in Northern Ireland. The chapter shows that the Agreement failed to change the identification of self and the other in the two communities. Protestants have continued to believe in the inherent superiority of the Protestant faith and in their right to rule. Catholics remain convinced that Protestant opposition to a united Ireland is based on the Protestant desire to maintain a position of privilege rather than any deep-rooted ideological conviction.

The third chapter in this section is by renowned American political scientist David Apter, who lived for a period among the Catholic community in the Bogside, Derry. It is a gripping account of how people on the ground have restructured their social lives around a politics of confrontation.

In the section on South Africa, chapters by Gagiano, Alexander, Adam and Schlemmer represent counterpoints which dramatically illustrate conflicting viewpoints of the current and future course of the struggle.

Drawing on a survey of white student attitudes, Gagiano argues that while whites of different political persuasions may disagree about the *government*, they are virtually unanimous in their support for the existing pro-capitalist and pro-middle-class *state*. The only structural reform they would countenance is one in which black representatives would share power in the government through strategic alliances with a predominantly white party, leaving the state intact.

Alexander, a radical black intellectual, questions whether such an alliance would ever be entertained by "revolutionary democrats" since it would perpetuate the historic connections between capitalism, apartheid and the state, albeit in a different guise. While the goal of the liberation struggle must remain the conquest of state power and socialist revolution, he argues that there is room for tactical alliances between reformists and revolutionaries which in the long run may help to transform civil society peacefully.

Adam and Schlemmer take up contrasting positions in reviewing the possibility of a compromise between representatives of the parliamentary and extra-parliamentary formations. Adam argues for the acceptance of the African National Congress as the patriotic force capable of creating one non-racial nation based on common citizenship and individual equality. In his view, Afrikaners and the wider white group would have to be persuaded that any racial form of nationalism is as unacceptable as colonialism and fascism.

Schlemmer, by contrast, warns against indulgence in utopian solutions which would suppress or eradicate deep-rooted feelings of political identity. He points to survey findings which suggest that a form of non-racialism resulting in an unqualified majoritarian outcome would be overwhelmingly rejected by whites and is in fact not demanded by most blacks.

In the last section of the book, the prospects for accommodation are assessed by Crick and Johnson. It is Crick's view that in divided societies, like the three under discussion, there is a strong tendency for individuals to find their human identity in being a member of a specific group. Accordingly, peace would require a compromise between communities.

Johnson points out that in South Africa, more than is the case in the other two societies, a compromise solution will have to be generated internally by the main adversaries, with little or no assistance from the outside.

Finally, Giliomee draws some conclusions from the proceedings of the conference and the individual contributions to the book.

ACKNOWLEDGEMENTS

The editors wish to thank the Friedrich Naumann Foundation and Idasa for contributing to the costs of the conference and of publishing the book, and Mobil Southern Africa for helping to defray some of the editorial expenses. In particular they would like to say thank you to the directors and staff at both Naumann and Idasa, and the participants in the Bonn Conference who wrote papers which for reasons of space could not be published in this book.

ENDNOTES

1. These distinctions are drawn from the essay by Theda Skocpol and Margaret Somers: "The Use of Comparative History in Macrosocial Inquiry", *Comparative Studies of History and Society*, 1980, pp174-97.

2. Stanley Greenberg: *Race and State in Capitalist Development: Comparative Perspectives*, New Haven, Yale University Press, 1980.

3. Heribert Adam: "Comparing Israel and South Africa: Prospects for Conflict Resolution in Ethnic States", unpublished paper.

4. Pierre van den Berghe: "South Africa and Israel as Herrenvolk Democracies", unpublished paper. George Frederickson in his *White Supremacy: A Comparative Study in American and South African History*, New York, Oxford University Press, 1981, also uses herrenvolk democracy as a leading theme in comparing the evolution of segregation in South Africa and the United States.

5. Sammy Smooha: "Control of Minorities in Israel and Northern Ireland", *Comparative Studies in Society and History*, 22, 2, 1980, pp258-59.

6. For an elaboration of this theme see Hermann Giliomee: "South Africa, Ulster, Israel: Elusive Search for Peace", *Optima* 36, 3, 1988, pp126-35.

7. Sammy Smooha: "Israel and the Palestinians in the West Bank and Gaza Strip: Is Partition Viable?", *Middle East Insight*, Fall 1988, pp28-35.

1 Ethno-nationalism and political instability: An overview

WALKER CONNOR

Looking backward

It risks triteness to note that during the past two decades ethno-nationalism has been an extremely consequential force throughout the First, Second, and Third worlds.[1] Even the more casual observers of the world scene now know that Belgium, France, Spain, and the United Kingdom are not ethnically homogeneous states and that the loyalty of Flemings, Corsicans, Basques and the Welsh to their respective states cannot be accepted as a given. Ethnic unrest within China, Romania, the Soviet Union, and Vietnam (to name but a few of the Marxist-Leninist states) is also a matter of record. Awareness of the significance of ethnic heterogeneity within the states of the Third World has reached the point where even newspaper accounts of coups, elections, and guerrilla struggles often contain references to the ethnic dimensions that are involved.

Few indeed are the scholars who can claim either to have anticipated this global upsurge in ethno-nationalism or to have recognised its early manifestations. With respect to the First World, there was a tendency to perceive the states as nation-states, rather than as multinational states, and, in any event, to presume that World War II had convinced the peoples of Western Europe that nationalism was too dangerous and outmoded a focus for the modern age. A supra-national, supra-state identity as European was perceived as the wave of the future. With regard to the Second World, it was broadly held that a highly effective power apparatus and the indoctrination of the masses in Marxist-Leninist ideology had made the issue of ethno-nationalism either superfluous or anachronistic. In effect, scholars accepted the official position of Marxist-Leninist governments that the application of Leninist national policy had solved "the national question", leading the masses to embrace proletarian internationalism. In Third World scholarship also, ethnic heterogeneity tended to be ignored or to be cavalierly dismissed as an ephemeral phenomenon. The catch-phrase of political development theory at the time was "nation-building", but its devotees offered few if any suggestions as to how a single national consciousness was to be forged among disparate ethnic elements.

How could this wide discrepancy between theory and reality be explained? More than a decade ago, this writer suggested twelve overlapping and reinforcing reasons:[2]

1. *Confusing interutilisation of the key terms.* Major result: a tendency to equate nationalism with loyalty to the state (patriotism) and therefore to presume that the state would win out in a test of loyalties.

2. *A misunderstanding of the nature of ethnic nationalism resulting in a tendency to underrate its emotional power.* Major result: a perception of ethnically inspired dissonance as predicted upon language, religion, customs, economic inequality, or some other tangible phenomenon, and, *propter hoc*, a failure to probe and appreciate the true nature and power of ethnic feelings.

3. *An unwarranted exaggeration of the influence of materialism upon human affairs.* Major result: the implicit or explicit presumption that the wellsprings of ethnic discord are economic and that an ethnic minority can be placated if its living standard is improving, both in real terms and relative to other segments of the state's population.

4. *Unquestioned acceptance of the assumption that greater contacts among groups lead to greater awareness of what groups have in common, rather than of what makes them distinct.* Major result: the optimistic belief that increased ties between groups are both symptomatic and productive of harmonious relations (as reflected, for example, in transaction-flow theory).

5. *Improper analogising from the experience of the United States.* Major result: a presumption that the history of acculturation and assimilation within an immigrant society would be apt to be repeated in multinational states.

6. *Improper analogising from the fact that improvements in communications and transportation help to dissolve regional identities to the conclusion that the same process will occur in situations involving two or more ethno-national peoples.*

Major result: a presumption that the waning of significance of regional identities in an ethnically homogeneous state, such as Germany, sets a precedent for multinational states.

7. *The assumption that assimilation is a one-directional process.* Major result: any evidence of a move toward acculturation/assimilation is viewed as an irreversible gain and is a basis for optimistic forecasts.

8. *Interpretation of the absence of ethnic strife as evidence of the presence of a single nation.* Major result: a tranquil period in the relations among two or more ethno-national groups causes scholars to assume that the society is ethnically homogeneous, or a multi-ethnic "national liberation movement" is perceived as mono-ethnic.

9. *Improper regard for the factor of chronological time and intervening events when analogising from assimilationist experience prior to the "Age of Nationalism".* Major result: examples of assimilation prior to the nineteenth century are employed as evidence that ethnic identity is a thoroughly fluid phenomenon.

10. *Improper regard for durative time by failing to consider that attempts to telescope "assimilationist time" by increasing the frequency and scope of contacts, may produce a negative response.* Major result: conviction that assimilation lends itself readily to social engineering.

11. *Confusing symptoms with causes.* Major result: explanations for political decay focus upon interim steps, such as the weakening of "mass parties", rather than upon the root cause of ethnic rivalry.

12. *The predisposition of the analyst.* Major result: a tendency to perceive trends deemed desirable as actually occurring.

The list is certainly not exhaustive, and at least five additional reasons suggest themselves:

13. *The mistaken belief that the states of Western Europe were fully integrated nation-states.* A number of leading theorists of political development explicitly maintained that the experiences of the states of Western Europe would be followed by those of the Third World in the course of their development. Thus, Western Europe was held up as an exemplar of something it was not, as proof that nation-states would develop in the Third World.

14. *A tendency to apply conventional scholarly approaches to the Third World.* A great deal of early Third World scholarship reflected the First World training of the analysts. Avenues of research, long applied to First World societies, were transferred to Third World states. Thus, a number of Third World studies explored political party structures and voting patterns in state assemblies, without appreciating that political parties were often viewed at the grassroots level as the continuation of ethnic rivalry by other means.

15. *Exclusive concentration upon the state.* Much of the early Third World scholarship reflected only the view from the capital, to the exclusion of the view from the ethnic homelands.

16. *Exclusive concentration on the dominant group in the case of societies with a "Staatvolk", such as Burma or Thailand.* One indication of this tendency can be found in purportedly state-wide political culture studies that make no reference to the political cultures of ethnic minorities, even when the latter account for a substantial percentage of the population.

17. *The tendency of many scholars to favour explanations based on class.*[3] Ethnic nationalism poses a severe paradox to such scholars since it posits that the vertical compartments that divide humanity into the English, Germans, Ibos, Malays, and the like constitute more potent foci of identity and loyalty than do the horizontal compartments known as classes.

Whatever the reason(s), the "nation-building" school failed to give proper heed to what, in most states, was *a,* if not *the,* major obstacle to political development. Today, just as two decades ago, ethnic nationalism poses the most serious threat to political stability in a host of states as geographically dispersed as Belgium, Burma, Ethiopia, Guyana, Malaysia, Nigeria, the Soviet Union, Sri Lanka, Yugoslavia, and Zimbabwe. Given, then, the failure of the political development theorists to reflect proper concern for the problems posed by ethnic heterogeneity, it is disturbing to still find no acknowledgement of this glaring weakness of the political development literature *by its most famous formulators.* Although numerous authors have drawn attention over the years to this remarkable slighting of the ethnic factor in the "nation-building" literature, the criticism has gone unanswered by those commonly identified with the fathering of political development theory.[4]

More recent developments

Scholarly indifference to problems arising from ethnic heterogeneity evaporated rapidly in the face of increasing numbers of ethno-national movements. By the mid-1970s, the study of ethnic heterogeneity and its consequences had become a growth industry. Literally thousands of articles focused principally on ethno-nationalism have appeared in English-language journals in the last decade. Scores of monographs and collections have been dedicated to the same topic, as have an impressive number of doctoral dissertations. Conferences on ethnicity have become commonplace, and panels on the subject have become regular parts of the programmes at annual meetings of professional organisations. A number of journals founded since 1970 — such as the *Canadian Review of Studies in Nationalism, Ethnic and Racial Studies*, and the *Journal of Ethnic Studies* — further attest to the intensified interest in ethnicity.

That this huge body of research has contributed magnificently to our knowledge of specific peoples, their inter-ethnic attitudes and behaviour, their leaders, and their aspirations is beyond dispute. But it must also be acknowledged that all of this scholarly activity has not produced an approximation of a coherent statement. There is a marked lack of consensus concerning how ethnic heterogeneity can be best accom- modated, or, indeed, even whether it can be accommodated non- coercively. And, in turn, these disagreements reflect a lack of consensus concerning the phenomenon that is supposedly the common focus of all these studies. In some cases, research, conducted under umbrella terms such as cultural pluralism, has in fact grouped several categories of identity (e.g., religious, linguistic, regional and ethno-national) as though they were one or, at least, as though they exerted the same impact upon behaviour. Others, influenced by the common misuse of the word ethnicity within the context of US society, have used this rubric when investigating nearly any type of minority found in a state (despite the fact that ethnicity was derived from the Greek word *ethnos*, connoting a group characterised by common descent). The analytical utility of such blanket categories is open to serious question. Minimally, such broad categories side-step raising the key question as to which of several group identities is apt to prove most potent in any test of loyalties.

Still other scholars, while focusing on ethno-nationalism, have described it in terms of some other "ism". Having already misassigned nationalism to loyalty to the state, they have perforce enlisted some other term to describe loyalty to one's ethno-national group. Primordialism(s), tribalism, regionalism, communalism, parochialism, and sub-nationalism are among the alternatives encountered most often. Each of these terms already had a meaning not associated with nationalism, a fact further contributing to the terminological confusion impeding the study of ethno-nationalism. (Communalism, for example, which refers within Western Europe to autonomy for local governments and within the Asian subcontinent to confessional identity, has appeared in the titles of books and articles in reference to ethno-nationalism in Africa and Southeast Asia.) Moreover, individually each of these terms exerts its uniquely baneful effect upon the perceptions of both the author and the reader. Collectively, this varied vocabulary risks misleading the reader into believing that what is misdescribed as regionalism within one country, tribalism within another, and communalism within a third are different phenomena — when in fact it is ethno-nationalism that in each case is the focus of the study. Imprecise vocabulary is both a symptom of and a contributor to a great deal of the haziness surrounding the study of ethno-nationalism. As noted elsewhere:

> In this Alice-in-Wonderland world in which nation usually means state, in which nation-state usually means multination state, in which nationalism usually means loyalty to the state, and in which ethnicity, primordialism, pluralism, tribalism, regionalism, communalism, parochialism and sub-nationalism usually mean loyalty to the nation, it should come as no surprise that the nature of nationalism remains essentially unprobed.[5]

Indeed, very few scholars have directly addressed the nature of the ethno-national bond. A common empirical approach of those who did so during the first half of this century was to ask the question: "What makes a nation?" Among the scholarly giants who raised this question were Carlton Hayes and Hans Kohn. They addressed themselves to what was necessary or unnecessary for a nation to exist. The typical response was a common language, a common religion, a common territory, and the like. Stalin's 1913 definition of a nation, which still exerts a massive influence upon Marxist-Leninist scholarship, was very much in this tradition:

> A nation is a historically evolved, stable community of people, formed on the basis of a common language, territory, economic

life, and psychological make-up manifested in a common culture.[6]

This approach still has its devotees, most notably Louis Snyder. Certainly it has the merit of emphasising the wisdom of employing a broadly comparative framework when studying ethno-nationalism. On the other hand, comparative analyses establish that no set of tangible characteristics is essential to the maintenance of national consciousness. Moreover, this particular approach would appear to fall into the trap mentioned earlier of mistaking the tangible symptoms of a nation for its essence. As the late Rupert Emerson cogently reminded us in a few prefatory words before launching an investigation of the elements that most commonly accompany national consciousness:

> The simplest statement that can be made about a nation is that it is a body of people who feel that they are a nation; and it may be that when all the fine-spun analysis is concluded this will be the ultimate statement as well. To advance beyond it, it is necessary to attempt to take the nation apart and to isolate for separate examination the forces and elements which appear to have been the most influential in bringing about a sense of the existence of a singularly important national "we" which is distinguished from all others who make up an alien "they". This is necessarily an overly mechanical process, for nationalism, like other profound emotions such as love and hate, is more than the sum of the parts which are susceptible of cold and rational analysis.[7]

As noted, very few of the present generation of scholars have attempted a serious probe of the nature of the ethno-national bond, forcing the curious to infer their conceptualisation of it from their comments concerning the causes or the solutions to ethno-national restlessness. When employing this yardstick, it appears that many authors have scant respect for the psychological and emotional hold that ethno-national identity has upon the group. To some, ethno-national identity seems little more than an epiphenomenon that becomes active as a result of relative economic deprivation and that will dissipate with greater egalitarianism. Others reduce it to the level of a pressure group that mobilises in order to compete for scarce resources. A variation on the pressure group concept places greater emphasis on the role of elites; rather than a somewhat spontaneous mass response to competition, the stirring of national consciousness is seen as a ploy utilised by aspiring elites in order to enhance their own status. Finally, in the hands of many adherents of the "internal colonialism" model, entire ethno-national groups are equated with a socio-economic class, and ethno-national consciousness becomes equated with class consciousness.

All of these approaches could be criticised as a continuing tendency of scholars to harbour what we termed earlier "an unwarranted exaggeration of the influence of materialism upon human affairs". They could also be criticised as examples of a tendency to misapply theoretic approaches (such as pressure group theory, elite theory and dependency theory). They can all be criticised empirically. (The most well-known propagator of the internal colonial thesis, Michael Hechter, has more recently recanted his support for that thesis, citing as cause its limited explanatory power.)[8] But they can all be faulted chiefly for their failure to reflect the emotional depth of ethno-national identity and the mass sacrifices that have been made in its name. Explanations of behaviour in terms of pressure groups, elite ambitions, and rational choice fail even to hint at the passions that motivate Kurdish, Tamil, and Tigre guerrillas or Basque, Corsican, Irish and Palestinian terrorists; or at the passions leading to the massacre of Bengalis by Assamese or Punjabis by Sikhs, or Tamils and Sinhalese by one another. In short, these explanations are a poor guide to ethno-nationally inspired behaviour.

Among the scholars who demonstrate a more profound regard for the psychological and emotional dimensions of ethno-national identity, there is a small but growing nucleus who, following the lead of Max Weber, now explicitly describe the ethno-national group as a kinship group. Among them are Joshua Fishman,[9] Donald Horowitz,[10] Charles Keyes,[11] Kian Kwan and Tomotshu Shibutani,[12] Anthony Smith,[13] and Pierre van den Berghe.[14] Interestingly, despite the fact that this formulation of the nation runs counter to classical Marxism, the Soviet Union's most influential academician in the study of national consciousness, Yu Bromley, also acknowledges the role of kinship in nation formation.[15]

Recognising the sense of common kinship that permeates the ethno-national bond clears a number of hurdles. First, it qualitatively distinguishes national consciousness from non-kinship identities (such as those based on religion or class) with which it has too often been grouped. Secondly, an intuitive sense of kindredness or extended family would explain why nations are endowed with a very special psychological dimension, an emotional dimension, not enjoyed by essentially functional or juridical groupings, such as socio-economic classes or states.

Unlike scholars, political leaders have long been sensitised to this sense of common ancestry and have blatantly appealed to it as a means of mobilising the masses. Consider Bismarck's famous exhortation to the German people, over the heads of their individual state leaders, to unite in a single state: "Germans, think with your blood!" Or consider the proclamation that was transmitted throughout fascist Italy in 1938:

The root of differences among peoples and nations is to be found in differences of race. If Italians differ from Frenchmen, Germans, Turks, Greeks, and so on, this is not just because they possess a different language and a different history, but because their racial development is different . . . *A pure "Italian race" is already in existence.* This pronouncement [rests] on the very pure blood tie that unites present-day Italians . . . This ancient purity of blood is the Italian nation's greatest title of nobility.[16]

Or listen to Mao Tse-tung in the same year describe the Chinese communists as "part of the great Chinese nation, flesh of its flesh and blood of its blood . . .".[17] Or read the programme of the Romanian communist party that describes the party's principal function as defending the national interest of "our people", a nation said to have been "born out of the fusion of the Dacians [an ancient people] with the Romans". Within Africa, Yoruba and Fang leaders have stressed a legend of common origin, as have Malay leaders within Malaysia.

It might well be asked why scholars have been so slow to discover what the masses have felt and what political leaders have recognised. There are several possible answers, including the intellectual's discomfort with the nonrational (note: *not* irrational), and the search for quantifiable and therefore tangible explanations. But another factor has been the propensity to ignore the vital distinction between fact and perceptions of fact. Several of the studies of the last generation to which we alluded did raise the issue of common ancestry as one of the possible criteria of nationhood. However, the authors then denied the relevance of such a consideration by noting that most national groups could be shown to be the variegated offspring of a number of peoples. But this conclusion ignored the old saw that it is not *what is*, but *what people believe is* that has behavioural consequences. And a subconscious belief in the group's separate origin and evolution is an essential ingredient of ethno-national psychology. A nation is a group of people characterised by a myth of common descent. Moreover, regardless of its roots, a nation must remain an essentially endogamous group in order to maintain its myth.

As noted, there are grounds for optimism in that a small core of influential scholars has come to recognise the myth of common ancestry as the defining characteristic of the nation. It may well be, therefore, that effective probing of the subjective dimensions of the national bond will occur during the next decade. Fishman has certainly begun excavating in this area, and Donald Horowitz indicates one avenue of possible fruitful research in suggesting that some of the work of experimental psychologists (dealing with both individual and group behaviour) may lead to a better understanding of ethno-nationalism. Moreover, Pierre van den Berghe now maintains that the literature on socio-biology has much to offer the student of ethnic identity.

Two other possibly productive areas of research for the probing of the emotional/psychological dimension of ethno-nationalism come to mind. The poet, as an adept articulator of deep-felt passions, is apt to be a far better guide here than the social scientist has proved to be. National poetry has hardly been touched upon on a worldwide comparative basis. Quite aside from aesthetics, it would obviously be of great value to learn what feelings and images have been most commonly invoked by recognised national poets, without regard to geography, level of their people's development, or other tangible distinctions.

Still another potentially fruitful source when probing the nature of ethno-nationalism consists of the speeches of national leaders and the pamphlets, programmes, and other documents of ethno-nationalist organisations. Too often these speeches and documents have been passed off as useless propaganda in which the authors do not really believe. But nationalism is a mass phenomenon and the degree to which the leaders are true believers does not affect its reality. The question is not the sincerity of the propagandist, but the nature of the mass instinct to which the propagandist appeals. Thus, Napoleon was unquestionably more a manipulator of than a believer in nationalism, but his armies were certainly filled with soldiers fired up by nationalism. Speeches and programmes should therefore be scanned from the viewpoint of comparative content. With what frequency do certain words and images appear? What referents are used to trigger the psychological response?

In any case, it is certain that very few scholars will attempt to probe the nature of nationalism. As in the past, most authors who touch on ethno-nationalism will be dealing with its manifestation in one or another society and/or policies aimed at its containment. However, their general treatment of the subject and their assessment of policies will necessarily reflect their unarticulated perception of the nature of ethno-nationalism, and it is to be hoped that the literature will increasingly embody that deeper respect for the emotional and psychological depths of ethno-nationalism that we have noted in the works of a small but growing number of influential writers.

Northern Ireland and South Africa

As conventionally described, the cases of Israel, Northern Ireland and South Africa would appear to have nothing in common. Only the Israeli situation has been broadly perceived by outsiders as an inter-ethnic conflict. The Northern Ireland problem has been popularly described as a religious struggle between Catholics and Protestants, and the South Africa issue as a racial confrontation between black and white. Such descriptions, however, mask the significant ethno-national dimensions of the Irish and South African cases.

Strife in the northern six counties of Ireland has been treated almost exclusively as a religious conflict, a quaint echo of the intra-national religious wars of a bygone age that saw Frenchmen pitted against Frenchmen, German against German, and so forth. To the degree that Northern Ireland's problem has not been viewed as religious, it has been treated as a civil rights struggle for political and economic reform. In fact, it is neither in its essence; rather, it is a struggle predicated upon fundamental differences in national identity. Contrary to the typical account, the people of Northern Ireland do not uniformly consider themselves Irish. Indeed, a survey conducted in 1968 by representatives of the University of Strathclyde indicates that a majority do not. Although 43 per cent of the respondents· thought of themselves as being Irish, 29 per cent considered themselves to be British, 21 per cent Ulster, and the remaining 7 per cent considered themselves to be of mixed, other, or uncertain nationality.

Unfortunately, the survey failed to correlate national identity with the religion of the respondents, but it is safe to assume, on the basis of the ethnic and religious histories of the island, that there exists a close correlation between self-identification as Irish and adherence to Catholicism. The important distinction, however, lies between those who consider themselves Irish, and those who either do not so consider themselves or are not so considered by the bulk of the Irish element. That the religious issue is largely extraneous helps to account for the fact that the consistent urging of tolerance by all but a handful of religious leaders has gone unheeded. Indeed, with at least as much accuracy, the conflict could be described as one of surnames rather than religions. Despite some intermarriage, the family name remains a fairly reliable index to Irish heritage, as compared to English or Scottish. It is for this reason that a surname is apt to trigger either a negative or positive response.[18]

In the case of South Africa, Hermann Giliomee purposefully and wisely eschews the word *ethnic* in favour of *communal*. As he notes:

> Ethnic is too narrow a category for it refers to a group with a common belief in a shared ancestry and history. This would fit Afrikaners, but not the larger white community, or, for that matter, the African or larger black community.[19]

However, as suggested by the present tricameral formal political struc- ture that provides separate legislative chambers to *whites, coloureds,* and *Asians* (thus slighting *blacks*), the situation is not a simple dichotomic one even at the communal level. The significance of the *coloured* and *Asian* components is suggested by a series of polls indicating that Anglophone whites are substantially better regarded among urban *blacks* than are *Asians* and *coloureds*. Conversely, asked in 1985 to name their preference for leader of the state, a majority (53,4 per cent) of Asians named then prime minister Botha and only 3,7 per cent named imprisoned African National Congress leader Nelson Mandela.[20] A majority of the Asian community would, therefore, apparently prefer to continue their present second-class status rather than risk the conse- quences of *black* rule. Consonant with this viewpoint, in a 1986 poll only six per cent of *coloureds* and five per cent of *Indians* (or *Asians*) looked favourably upon the prospect of a "black majority government" as compared with 31 per cent and 54 per cent (respectively) who looked favourably upon a "white-dominated government".[21]

The multiplicity of communal groups and intercommunal attitudes are therefore significant to an understanding of the South African scene. More important to our particular concern with comparability, however, is the ethnic dimension that is simultaneously in operation. That the Afrikaners constitute an ethno-national group is beyond doubt. And although relations between them and the non-Afrikaner *white* community have significantly improved during the last decade, it would be in error, as earlier suggested,[22] to interpret a period of ethnic tranquility as evidence of the absence of diverse groups.[23] Similarly, while it is currently bad form to draw attention to the ethnic divisions among *blacks* because of the manipulative nurturing to which they have been subjected by the government, those divisions are quite real. Thus, an attitudinal survey of urban Zulus indicated that their strongest dislike was felt, in descending order, toward *Asians, coloureds,* and *whites* (otherwise undifferentiated), but a significant level of distaste also characterised their attitudes toward other black ethnic groups (Ndebele,

Tsonga, Tswana, Xhosa, Sotho, and Swazi).[24] For example, some 30 per cent of respondents would not even consider marrying an Ndebele or Tsonga.

These divisions are clearly not without ramifications for the country's power struggle. In his running battle with the ANC, Zulu leader Buthelezi has stressed that the ANC is Xhosa dominated. And the close association between the ethnic map and the likelihood of conducting successful guerrilla warfare within South Africa has been underscored by another insurrectionist group, the Pan-Africanist Congress of Azania (PAC), in a critique of ANC operations:

> Problems of language, terrain, and the cardinal requirement of mass support make it difficult, if not impossible, for guerillas to operate in territories from which they do not originate.[25]

Acknowledging that each of the three societies — Israel, Northern Ireland, and South Africa — contains ethno-national groups is not to say that there is no ambiguity concerning group identity; witness the non-Irish element within Northern Ireland. These are people principally of Scottish Lowlander or English descent, many of whose ancestors came to Ireland nearly four hundred years ago. With a sense of Scottish or English heritage grown dim by centuries of separation from the parent group, these people have had trouble defining themselves in ethnic terms. The frustration wrought by this confusion was captured in an interview with Glenn Barr, a member of the non-Irish community: "I don't know what I am. People say I'm British. The British treat me as a second-class citizen. I am not Irish. I am an Ulsterman."[26]

The tendency to define themselves in negative terms, that is, *not* Irish, is reflected in the decreasing numbers willing to use the term *Irish* when identifying themselves.

Table 1.1: Identity poll data[27]

	British	Ulster	Irish	Other
1968	39	32	20	9
1978	67	20	8	5
1985	60,7	34,8	4,6	—

As with the description *Ulster*, it is likely that the term *Irish* is perceived by those non-ethnically Irish who select it as a geographic rather than an ethnic expression and that its decreasing popularity is a reflection of the growing tendency to recognise the competition as ethno-national, with a corresponding drop in the use of Protestant and Catholic to define the contending groups. Something of the sort occurred amongst the ancestors of these people who migrated from Ireland to the United States. At first they favoured the term *Irish* to describe themselves, but, as ethnically Irish migrants began to arrive in numbers, *Scotch-Irish* became the favoured self-description as a means of differentiating themselves both from the Scots of Scotland and the Irish.[28]

The lack of interest within Northern Ireland in being identified as Scottish or English suggests that what has been termed an *offshoot* nation may be evolving.[29] Although they have not yet agreed upon a descriptive term, have this people been following an Afrikaner-style path toward a sense of separate nationhood?[30]

Identity in the Israeli situation is also not without ambiguity. Two decades ago, the Ashkenazim-Sephardim division of the Jewish community, reinforced by often discernible physical differences,[31] was considered by some to be of enduring significance for ethnic identity.[32] However, as a result of the ability of a myth of common identity to overcome contrary fact, this division is rapidly losing impact.[33]

More significant is the issue of whether the more fundamental identity of the non-Jewish population is Palestinian or Arab. Assertions concerning both a Palestinian nationalism and an Arab nationalism are common. In some cases, an author has attributed both an Arab nationalism and a Palestinian nationalism to the same people without feeling compelled to offer any further explanation. Recently the issue has become part of a larger question concerning whether primary identity throughout Arabdom as a whole is focusing on Arabness or on any one of the eighteen states into which the "Arab world" is divided. The literature has been definitely weighted in favour of state (Iraqi, Kuwaiti, Moroccan, and so on) identity, but the supporting data and the interpretation of that data leave much to be desired.[34]

The comparability of Israel, Northern Ireland and South Africa

No matter how carefully cases are selected for comparative analysis, it is essential not to lose sight of the fact that the selections are only imperfectly analogous. The validity of inferences drawn from the comparison will depend as much upon sensitivity to dissimilarities as upon appreciation of similarities. A comparison of the situations within Israel, Northern Ireland and South Africa might therefore beneficially be prefaced by considering a few of the major differences and commonalities.

Here we are primarily interested in factors which contribute to the peculiar mind-set of the politically dominant group in each of the three settings. To recite the obvious, the possibility of achieving a peaceful accommodation will depend to a great degree not on an objective analysis of the situation, but on how the parties subjectively view it.

The relative numerical size of the politically dominant group is of great importance. The fact that Jews within either pre- or post-1967 Israel and the non-Irish community within Northern Ireland each represent a safe majority has permitted them to draw upon principles of democratic legitimacy to justify policy. Whether or not conventional democratic principles and institutions are appropriate to a multi-national setting is at least questionable. As noted elsewhere:

Failure to appreciate both the hierarchical nature of group loyalties and the emotional depth of nationalism is reflected in one school of thought, holding that the fact that people simultaneously partake of several group identities is itself an inbuilt guarantee that nationalism can be accommodated in a multinational, democratic state. The approach is reminiscent of James Madison's in *Federalist Papers* Number 10, in which he described the manner in which cross-cutting interests naturally balance one another. Starting from the fact that the economic, ethno-national, racial, religious, sectional, social, and other group identities of an individual are not coterminous, this school contends that the overlapping interests arising from all group identities will act as a check against any single identity leading to a desire to withdraw from political society. In effect, this same presumption undergirds the democratic doctrine of the changing majority. Though a person may be in a minority when matters involving one of his identities are at issue (for example, a religious issue), he can expect to be in a majority when the principal issue of the day involves another form of identity (perhaps his social class). An important assumption of this doctrine is that any majority will show great restraint and a willingness to accommodate the minority, because its members realise that they may be in the minority tomorrow. But what happens if the emotional attachment to various identities is not of the same importance, if one is so powerful that it takes precedence over all others? History's many nationally inspired separatist movements indicate that greater significance has been popularly ascribed to ethno-national divisions than to the aggregate of commonalities that bridge that division. As a result of the preeminence ascribed to national identity, the political system of an ostensibly democratic society such as Northern Ireland may be accurately defined as the permanent majority. In such case, democratic institutions become largely facade for perpetuating domination of one group by another. In sum, given the hierarchical nature of group commitments, it is questionable whether one person/one vote democracy can operate as designed in a multinational state.[35]

Be that as it may, the democratic myth which the numerical advantage of Jews and "Ulstermen" has permitted has not only been a major element in seeking sympathy and support abroad but has been a major factor of reinforcement in the conviction of members of these dominant groups of the unquestionable righteousness of their actions.

By contrast, the fact that the Afrikaners account for only some 10 per cent of South Africa's total population and that even the consolidated *white* community accounts for less than one-fifth of the population not only helps to explain South Africa's far more isolated and reviled status relative to that of Israel and Northern Ireland, but also the increasing self-questioning of their right to paramountcy by members of the dominant group.[36] For example, while some 60 per cent of Afrikaners now accept power-sharing as "unavoidable", less than 16 per cent of the Jewish public believe that Arabs should be treated equally within Israel.[37] "Ulsterman" satisfaction with the present power structure in Northern Ireland is suggested by the fact that 90 per cent perceive the police as being fair and 89 per cent believe that the legal system dispenses justice fairly.[38] Indeed, the major policies of the South African government since 1948 — apartheid, bantustans, attempts to drive wedges between the various ethnic groups, and the more recent pattern of searching for formulae that would grant (albeit grudgingly) additional social and political rights to *blacks* — can all be explained as a means of mitigating the psychological effects of numbers.

The numerical disadvantage of the Afrikaners also helps to explain the prevalence of attitudinal and behavioural patterns popularly referred to collectively as a siege mentality. However, this outlook does not vitally distinguish Afrikaners from "Ulstermen" and Israeli Jews. Ulsterite and Jewish perceptions of their situation cannot be divorced from the larger Irish and Arab communities across the border. The same majority in one context becomes a threatened minority in another. Siege mentality is therefore more of a commonality than a dissimilarity.

One major factor that differentiates Northern Ireland from the other two cases flows from the fact that it is not a sovereign state. Major decisions are made outside of the region and outside of the dominant group. Many actions taken in the name of either the Ulster or Irish community are ultimately aimed at the populace of Great Britain rather than at one another.

A common element among the cases is that the politically dominant group is perceived as a settler community. Sensitive to this characterisation with its overtones of imperialism and illegitimacy, the Israelis stress that their claim dates to the period of David (1000–961 BC) and beyond; the Ulstermen emphasise the multi-century length of their presence (which, it is often pointed out, approximates the uninterrupted presence of Europeans in North America), and the Afrikaners stress both their multicentury presence and the charge that most of the present black inhabitants of South Africa arrived from the north still later. The point, in any case, is that the perceptions and counter-perceptions concerning indigeneity are important elements in all three cases.

The accommodation of heterogeneity

Questions of accommodating ethno-national heterogeneity within a single state revolve about two loyalties, loyalty to the nation and loyalty to the state, and the relative strength of the two. The great number of bloody ethno-national movements that have occurred in the past two decades within the First, Second, and Third worlds bear ample testimony that when the two loyalties are seen as being in irreconcilable conflict, loyalty to the state loses out. But the two need not be so perceived. To people with their own nation-state or to those people who are so dominant within a multinational state as to perceive the state as essentially their nation's state (for example, the English, the Han Chinese, the Thais), the two loyalties become an indistinguishable, reinforcing blur. It is in the perceptions of national minorities that the two loyalties are most apt to vie.

During the last fifteen years, scholarship has made some important strides in probing the attitudes of minorities toward the state, although much more remains to be done. A number of sophisticated analyses have made good use of attitudinal data to bring us beyond the stage of simply hypothesising about the two loyalties.[39] It should be noted, however, that the literature is based overwhelmingly, although not exclusively, on First World states and on homeland peoples.

The following findings emerge from these studies:

1. Members of ethno-national minorities manifest substantially less affection toward the state than do members of the dominant group.
2. Minorities of the same state can differ significantly in this regard.
3. For most persons, however, the matter is not perceived in either/or terms. Affective ties to the state coexist with ethno-national consciousness.
4. In most cases in which a separatist movement is active, large numbers, usually a majority of the involved group, do not favour secession.
5. In some cases, the percentage represented by those in favour of secession has remained relatively constant; in other cases it has evidenced profound trends.
6. Regardless of their attitude toward secession, a preponderant number do favour major alterations in the political system that would result in greater autonomy.
7. Where separatist parties are allowed to contest elections, their vote is not an adequate index to separatist sentiment.
8. In all cases for which there are attitudinal data, members of ethno-national groups overwhelmingly reject the use of violence carried out in the name of the national group.
9. However, a large percentage, including many who do not favour separation, empathise with those engaged in violence and place the blame for the violence upon others.
10. Separatists draw their support from all social strata and age groups.
11. Disproportionate support, however, comes from those under 35 years of age, with above-average education and income.
12. Professional people are disproportionately represented.
13. Lack of support is particularly pronounced among those over 55 years of age.

Many of the preceding points probably appear trivial or trite. Given the numerous powers at the disposal of the state for politically socialising its population, it is hardly surprising to find that affective ties to the state exist among minorities. However, it would be particularly dangerous to take this finding, predicated principally upon First World states, and apply it wholesale to a Third World environment. For one thing, most Third World states are too young to have developed the sense of institutional and symbolic legitimacy that is a central aspect of state loyalty. And the older Third World states (such as Afghanistan, Ethiopia, Iran, Liberia, and Thailand), given the historic absence of the principal means for inculcating this sense of legitimacy (for example, a public school system), are not true exceptions. To take perhaps an extreme case, it would be foolhardy to presume that the Kachins, Karens, and Shans harbour any noteworthy level of goodwill toward the Burman state.

The finding that most members of national minorities are prepared to settle for something less than separation probably has more universal application. The still revolutionary idea popularly termed national self-determination holds that any people, simply because it considers itself to be a separate people, has the right, *if it so desires*, to create its own state. However, it would appear to be the rule that a majority of members of a homeland people are prepared to settle for autonomy for the homeland. Even when demands are made for actual separation, Third World elites are usually as fragmented as those in the First World — between those who maintain this stance and those who announce their willingness to settle for autonomy. (The Pakistani Baluch, the Iraqui Kurds, the Moros and the Sikhs are major current illustrations.) Moreover, a number of groups, although having engaged in violent struggle for the stated aim of independence, have subsequently entered into a peaceful relationship with the state authorities on the basis of a grant of autonomy. In the typical pattern, these periods of peace disintegrate amidst a flurry of charges and countercharges over whether the government's promises of autonomy have been honoured. Underlying these failures at accommodation have usually been differing views concerning the content of autonomy.

Although often treated by scholars, state authorities and ethno-national elites as alternatives, independence and autonomy are hardly that. Autonomy is an amorphous concept, capable of covering a multitude of visions extending from very limited local options to complete control over everything other than foreign policy. It can therefore incorporate all situations between total subordination to the centre and total independence. Both autonomy and independence are therefore terms that tend to obscure important shadings in the attitudes that members of a group can be expected to hold concerning goals.

It should not be surprising that ethno-national peoples should blur the distinction between independence and autonomy. Ethno-national concerns, by their very nature, are more obsessed with a vision of *freedom from* domination by non-members than with a vision of *freedom to* conduct foreign relations with states. They are the reaction to international, not interstate, relations. The average Basque or Fleming, just as the average Kurd or Naga, does not appear to be influenced by a prospect of a seat at the United Nations or an embassy in Moscow. Indeed, in the case of most Third World peoples, meaningful autonomy would represent a return toward either the loose system of feudatory allegiances they knew under Afghani, Chinese, Ethiopian, or Persian empires or to the indirect rule that they knew under colonialism, both of which had more effectively muted ethno-national concerns than have their successor political systems. In short, ethnocracy does not presuppose state independence. It does presuppose meaningful autonomy.

The finding concerning popular sympathy for those who carry out violence in the name of the national group also has momentous implications for the political stability of states, for it explains how guerrilla struggles have been maintained for years in the face of overwhelming odds. It undergirds the wisdom of Giuseppi Mazzini's statement of more than a century and a half ago: "Insurrection — by means of guerilla bands — is the true method of warfare for all nations desirous of emancipating themselves from a foreign yoke." In the case of wars of (ethno) national liberation, the numbers actually engaged in guerrilla struggle may be quite small, but those who fight in the name of the nation's liberation can expect that degree of sympathy that, as Mao Tse-tung, Truong Chinh and others have noted, is indispensable to the conduct of a successful guerrilla struggle. And this is why a number of leaders of guerrilla struggles, whose own goals have had nothing to do with minority rights, have gained the necessary local support by promising independence or autonomy to ethno-national groups. Such promises played significant roles in the assumption of power by the Chinese communist party, the Viet Minh and the Pathet Lao, and it is today a key element in the propaganda of revolutionary movements in Latin America. In Africa also, the numerous guerrilla struggles cannot be understood without reference to ethnic maps and aspirations.

While the thirteen findings outlined above attest to the formidable threat that ethnic heterogeneity poses to political stability, they also contain much to encourage those in search of formulae for accommodating such diversity. The fact that most members of most ethnic minorities are prepared to settle for something less than complete independence means that such formulae are not sheer whimsy. The likelihood of arriving at a mutually agreeable formula is another matter, however. A successful formula will require a significant measure of decentralisation of authority, and governments, by nature, are ill-disposed towards the relinquishing of power. However, governments may come to recognise that a measure of devolution would actually increase their authority over otherwise rebellious national groups. Few students of Spanish politics prior to 1975 imagined that the highly centralised, authoritarian system of Franco would be followed by a government prepared to grant substantial autonomy to the country's non-Castillian peoples. Belgium, Canada, and Panama are other states that have recently diluted support for separatist movements by adopting power-sharing formulae. Most governments, however, have not been prepared to grant the degree of autonomy necessary to avoid the resort to separatism. But the tendency for ethno-national groups to aspire to greater autonomy, while being prepared to settle for something short of full independence, does underline the fact that a solution to ethnic heterogeneity must ultimately be found in the political sphere and not in the economic one.

Summary

The literature on ethno-national heterogeneity has undergone a quantum leap in the last decade. Our knowledge of specific peoples and problems has grown enormously as a result. This outpouring, however, has not resulted in a broad consensus concerning either the nature of ethno-nationalism or means to its accommodation, although there are grounds for optimism that substantive progress will be made during the next decade.

My own admonitions for ensuring this progress are five in number:

1. Greater attention must be paid to avoiding imprecise and confusing terminology.[40]
2. Greater appreciation for the psychological/emotional depth of ethno-national identity must be reflected in the literature.
3. Greater refinements are necessary with regard to classifying peoples and political systems for comparative purposes.
4. Greater appreciation that ethno-national demands are at bottom political rather than economic in nature should be reflected in proposals for accommodating ethnic heterogeneity.[41]
5. It should always be remembered that ethno-nationalism is a mass phenomenon, and keeping this in mind should counteract the tendency to overemphasise the role of elites as its impresarios.[42]

ENDNOTES

1. The following article draws heavily from the author's "Ethno-nationalism" in Myron Weiner and Samuel Huntingdon (eds): *Understanding Political Development*, Boston, Little, Brown and Company, 1987, pp196-220.

2. For a lengthier discussion and illustrations of each of these items, see Walker Connor: "Nation Building or Nation-destroying?", *World Politics* 24, No 3, April 1972, pp319-55.

3. I am indebted to Myron Weiner for drawing attention to this omission from my list.

4. Thus, a recent article by Gabriel Almond, intended as a retrospective on the political development literature and its critics, makes no reference to ethnic heterogeneity nor to its slighting by the political development school. See "The Development of Political Development" in Myron Weiner and Samuel Huntington (eds): *Understanding Political Development*, Boston, Little, Brown and Company, 1987, pp437-90.

5. Walker Connor: "A Nation is a Nation, Is a State, Is an Ethnic Group, Is a . . .", *Ethnic and Racial Studies* No 4, October 1978, p396.

6. Joseph Stalin: *Marxism and the National Question*, Moscow, Foreign Languages Publishing House, 1950, p6.

7. Rupert Emerson: *From Empire to Nation: The Rise to Self-assertion of Asian and African Peoples*, Cambridge, Mass, Harvard University Press, 1960, p102.

8. Michael Hechter, Debra Friedman and Malka Appelbaum: "A Theory of Ethnic Collective Action", *International Migration Review* 16, No 2, Summer 1982, pp412-34. The article stresses rational choice as a means of overcoming the explanatory shortcomings of group stratification.

9. Joshua Fishman: *The Rise and Fall of the Ethnic Revival in the USA*, The Hague, Mouton, 1985.

10. Donald Horowitz: *Ethnic Groups in Conflict*, Berkeley, University of California Press, 1985.

11. Charles F Keyes: "Towards a New Formulation of the Concept of Ethnic Group", *Ethnicity* 3, No 3, September 1976, pp202-13.

12. Kian Kwan and Tomotshu Shibutani: *Ethnic Stratification: A Comparative Approach*, New York, Macmillan, 1965, p47.

13. Anthony D Smith: *The Ethnic Revival*, Cambridge, Cambridge University Press, 1981.

14. Pierre van den Berghe: "Race and Ethnicity: A Sociobiological Perspective", *Ethnic and Racial Studies* 1, No 4, October 1978, pp401-11.

15. Yu Bromley: "Ethnography and Ethnic Processes", *Problems of the Contemporary World* No 73, Moscow, USSR Academy of Sciences, 1978.

16. "Manifesto of the Racial Scientists (July 14, 1938)", reprinted in *Mediterranean Fascism: 1919-1945*, edited by Charles Delzell, New York, Harper and Row, 1971, pp174-5.

17. *Selected Works of Mao Tse-tung*, Peking, Foreign Languages Press, Vol 2, 1975, p209.

18. Connor: "Nation Building", pp339-40. See also Walker Connor, "Ethnonationalism in the First World: The Present in Historical Perspective" in Milton Esman (ed): *Ethnic Pluralism and Conflict in the Western World*, Ithaca, NY, Cornell University Press, 1979, pp40-1. For an example of a scholarly paper that reduces the conflict to a religious struggle, see A S Cohan: "The Question of a United Ireland: Perspectives of the Irish Political Elite", *International Affairs* 53, April 1977, p242: "The cultural divide in Northern Ireland is a religious one. . . . The critical fact in the separation of the two Irelands is found in the religious dimension." By contrast, a French photographer, who has covered the struggle for two decades, recently stated in an interview that he no longer perceived it as a religious conflict. He is quoted as stating "it's about nationhood". This insight appears to have exerted little impact upon the reporter, however, who opens his article: "Gilles Peress, a French photographer, arrived in Belfast on July 11, 1970, with . . . an almost total ignorance of the Irish war of religion." (See Cal McCrystal: "Twenty Years of Torment", *The Sunday Times Magazine*, London, 13 August 1989, p18.)

19. "The Communal Nature of the South African Conflict" in Hermann Giliomee and

Lawrence Schlemmer (eds): *Negotiating South Africa's Future*, Johannesburg, Southern, 1989, p114.

20. By contrast, 54,2 per cent of the *black* sample named Mandela.

21. Cited in Lawrence Schlemmer: "Processes and Strategic Options for South Africa", *SAIS Review* 9, Winter/Spring 1989, p115.

22. See above, p4, item number 8.

23. For excellent survey data documenting a continuing and growing gap between the Afrikaner and non-Afrikaner *white* communities and their respective attitudes toward the state, see the contribution to this volume by Jannie Gagiano.

24. Brian du Toit: "Ethnicity, Neighbourliness and Friendship among Urban Africans in South Africa", in Brian du Toit (ed): *Ethnicity in Modern Africa*, Boulder, Westview Press, 1978, pp143-74. See also the self-identity data on the part of *blacks* in the contribution to this volume by Hermann Giliomee.

25. Walker Connor: "The Political Ramifications of Ethnic Diversity within Subsahara Africa", unpublished paper, 47.

26. *New York Times*, 16 November 1987.

27. The polling data for 1968 and 1978 were collected by Richard Rose and E Moxon-Browne, respectively, and are cited in Roy Wallis, Steve Bruce and David Taylor: "Ethnicity and Evangelicalism: Politics in Ulster", *Comparative Studies in Sociology and History* 7, April 1987, p301. The 1985 data are drawn from Desmond Bell: "Acts of Union: Youth Sub-culture and Ethnic Identity amongst Protestants in Northern Ireland", *British Journal of Sociology* 38, June 1987, p179.

28. *Harvard Encyclopaedia of American Ethnic Groups*, Cambridge, MA, Harvard University Press, 1980, p906.

29. Offshoot nations are formed when an important segment of a nation has been geographically separated from the parent group for a period of time sufficient for it to develop a strong sense of separate consciousness. Members retain an awareness that they derive from the parent stock, but they believe that the characteristics they have in common are less significant than those that make them unique. Examples include the Afrikaners, the Formosan Han-jen and the Quebecois.

30. The popularity of the political or citizenship-indicating term *British* is probably not too significant in this regard. To most it unquestionably reflects the desire to remain part of the United Kingdom rather than risk absorption by the Republic of Ireland. Should Britain decide to withdraw from Northern Ireland, a substantial decrease in the number identifying themselves as *British* could be expected. The ancestral division of the community between English and Scottish should pose less of a problem to the creating of a single identity than would have been expected of the Dutch and French Huguenot backgrounds of the Afrikaner, because the Scottish ancestors were Lowlanders and therefore of the same Anglo-Saxon-Norman background as the English.

31. In polls directed by Yochanon Peres, which divided interviewees into Ashkenazi, Sephardic and Arab, the interviewer was instructed in the case of a Jew, to secretly note whether the respondent appeared to be Ashkenazi or Sephardic. See, for example, his "Ethnic Relations in Israel" in Michael Curtis (ed): *People and Politics in the Middle East*, New Brunswick, NJ, Transaction Books, 1971, pp31-68.

32. Maxime Robinson is but one of many authors who noted that many who are today termed Palestinian Arabs were related historically to the Sephardic Jews. See *Israel and the Arabs*, Baltimore, Penguin Books, 1968, pp213-14.

33. In a recent, excellent polling study of social distance between Jews and Arabs, Sammy Smooha: "Jewish and Arab Ethnocentrism in Israel", *Ethnic and Racial Studies* 10, January 1987, pp1-26, did not feel it necessary to disaggregate the Jewish responses into Ashkenazi and Sephardic components.

34. See particluarly, Tawfic Farah and Yasumasa Kuroda (eds): *Political Socialisation in the Arab States*, Boulder, Lynne Rienner Publishers, 1987, and Tawfic Farah (ed):

Pan-Arabism and Arab Nationalism: The Continuing Debate, Boulder, Westview Press, 1987. The new emphasis on state-identity appears to have been sparked by an article by Fouad Ajami: "End of Pan-Arabism", *Foreign Affairs* 57, Winter 1978/79, pp355-73.

35. Walker Connor: "Nationalism: Competitors and Allies", *Canadian Review of Studies in Nationalism* X, Fall 1983, p279.

36. The Zulus and Xhosas are each numerically greater than the combined *white* community.

37. Lawrence Schlemmer: "Processes . . ." p116 and Smooha: "Jewish and Arab" p13.

38. Ronald Weitzer: "Contested Order: The Struggle over British Security Policy in Northern Ireland", *Comparative Politics* 19, April 1987, p286.

39. Particularly noteworthy has been the work of Maurice Pinard on Canada and Robert Clark on Spain.

40. Note, for example, in the essay by Gabriel Almond (referred to above in 4), the use of "the nation-state" to describe all states and his use of "nations" to describe the states of the Middle East, Africa and Asia. The continuing misuse of these key terms certainly reflects and may very well help to explain Almond's failure to confront the problems that ethnic heterogeneity poses for political development.

41. For a more detailed discussion, see Walker Connor: "Eco- or Ethno-nationalism?", *Ethnic and Racial Studies* 7, No 3, July 1984, pp342-59.

42. As set forth by a scholar more than forty years ago: "A history of national consciousness should not, like a history of philosophy, simply describe the thought of a limited number of eminent men without regard to the extent of their following. As in the histories of religions, we need to know what response the masses have given to different doctrines." Walter Sulzbach: *National Consciousness*, Washington DC, American Council on Public Affairs, 1943, p14.

2 The dominant communities and the costs of legitimacy

MICHAEL MACDONALD

*"The strongest is never strong enough to be always the master,
unless he transforms strength into right and obedience into duty."*
– Jean Jacques Rousseau

Politics in Northern Ireland, South Africa and Israel is characterised by chronic and intense conflict, punctuated with gruesome and often indiscriminate violence. It is often hoped that the sufferings borne by members of the various communities in these societies might foster support for moderate political changes, changes that would recognise the legitimacy of the political aspirations of the interested communities and represent mutually beneficial and acceptable accommodations among the conflicting traditions. The purpose of this chapter is to explain why this hope meets powerful opposition, why intransigence, polarisation, and violence often prevail over political compromise in Northern Ireland, South Africa and Israel, and why, in spite of the obvious and manifest differences between the situations and their conflicts, none of the three societies has reached, or even *approached,* a political settlement among its diverse communities.

The thesis is not that political change is impossible in Northern Ireland, South Africa or Israel; it is, instead, that the costs to the dominant community of accepting the legitimacy of the political aspirations of the weaker community are higher than is often supposed, that the costs flow from the centrality of political power to both the socio-economic interests and the collective self-identities of substantial sections of Northern Irish Protestants, white South Africans and Israeli Jews, and that the nature of these interests and identities reduces the prospects for voluntary and consensual political changes in all three societies.

It is often noted that economically marginal and vulnerable Northern

Irish Protestants, South African whites and Israeli Jews are especially hostile to the political demands of, respectively, Northern Irish Catholics, South African blacks and Palestinians. It is also argued that economic interests, especially in South Africa, exert pressure for a stable political order, which would allow for economic growth without the costs of political disruption. The idea is that, while some members of the dominant communities in each society might depend on special protections to maintain their economic positions, most economic interests would prefer stability and prosperity to instability and destruction.

This position has much to recommend itself, but it does miss the implications of the fact that in all three societies the economic interests of influential members of the dominant communities, and not only the lower strata, *derive* from, and *depend* on, political power. The economic interests of influential constituencies — including state bureaucrats, employees and managers of parastatal enterprises, the subsidised middle classes, and protected workers and supervisors — are grounded in access to state resources, which means that they presume more political influence for themselves and, by implication, less influence for the excluded and exploited economic interests. Their interests are not prior to, and separable from, the political statuses of the politically empowered communities, but rather presume political access for the beneficiaries of state patronage. The political liberalism of the dominant communities is restricted, therefore, by the weakness of economic liberalism; public political interests permeate private economic ones.

The political empowerment of one community and the dis-empowerment of the other predicates communal identities as well as economic interests. The identities of the dominant communities, particularly Northern Irish Protestants and South African whites, do not just sprout spontaneously over the course of history, but are constructed by political means and through political control. Note, for example, that the categories that identify these groups — "Protestants" and "whites" — are comprised of otherwise disparate, even acrimonious elements —English Anglicans versus Scottish Presbyterians in the one case, English-speakers versus Afrikaners in the other.

This results from the fact that the unity of Northern Irish Protestants and South African whites is not, at the outset, cultural or ethnic. Instead, it is constituted by the full political membership that these people share and that the other groups — Irish Catholics and South African blacks[1] — lack. Certainly the situation is different for Israeli Jews, since Jews have existed as a cohesive and identifiable people for thousands of years prior to the establishment of Zionism and the foundation of modern Israel. But even in this case the threat posed by, and the exclusion of, Palestinians in the occupied territories helps to mediate differences among Israeli Jews, particularly by assisting Sephardic Jews with the otherwise difficult project of identifying with a generally Ashkenazi state.

Thus Northern Irish Protestants, South African whites and (to a lesser degree) Israeli Jews are constituted as communities both by shared internal characteristics and by contrast with the other, subordinate communities. What matters is not just what they share, but what they share and what the other lacks, which is, above all else, meaningful political membership.

The centrality of political power to the socio-economic interests and the collective traditions of Northern Irish Protestants, South African whites and Israeli Jews expresses the comparable — not to say identical — origins of all three societies. The original similarity among the three cases is the act of settlement, which not only brought the respective sides — Protestant and Catholic, white and black, Jew and Palestinian — into contact in the first place, but which, together with the experience of British imperialism, also constituted all of these otherwise disparate groups into more coherent and cohesive communities. What inaugurated the conflicts was that the settlers settled and that their descendants stayed, that the new peoples planted themselves and took root in their new land.

The tragedy of the situation is that it was the depth of the settlers' commitment to their homes, and not the superficiality, that worsened relations with the indigenous population. If the original settlers had just used and discarded the colony in the fashion of conventional imperialists, or practised genocide in the fashion of white Americans or Australians, the conflicts with the native population would, ironically, have been less severe.

In the case of genocide, the settlers eliminate the prospect of future conflict; and in the case of conventional imperialism, the settlers were content to skim what they could from their investments and then depart, not bothering to reshape the colony in their own image and not locking themselves into mortal conflicts with natives. Their incursions disrupted the integrity of the traditional societies, of course, but colonialists motivated by simple rapacity were not dedicated to effacing what had preceded them.

Northern Ireland, South Africa and Israel reflect their origins as settler societies. In each case, the descendants of the settlers have resided continuously in their current home for generations; and in the case of Israel, the claim is made that the Jews "returned" to their historic lands, rather than settled those of another people. Accordingly, Northern Ireland, South Africa and Israel are divided societies, but of a distinctive sort, ones in which social, political, and economic relations are built on, and reproduce, the prior clashes between the settling and the indigenous communities.

Power relations between the communities are fundamentally and persistently unequal. The interests, identities and traditions of both the dominant and subordinate communities develop through *interaction with* their opposites and not side-by-side and separately. Important sections of the superordinate groups identify themselves with their superiority *as* superiority and come to depend for their self-definition not only on what they have, but on the discrepancy between what they have and what the subordinate group lacks. The meanings assigned to the different origins of the settling and the indigenous communities become the grounds for dichotomous social and political institutions.

Whether such societies ought to be classified as settler societies by virtue of their origins and the continuing conflicts with the descendants of the indigenous population, or whether the commitment of the descendants of the early settlers to their new homes and their hostility to their erstwhile patron countries render these as something other than settler societies, is not at issue here. Either way, the prevailing institutions elaborate on rather than mend the original contradictions between the antagonistic populations. The identifications that distinguish one community from the other are expressed first in political, social and economic institutions and eventually are converted into something akin to ontological statuses.[2]

The political aspirations of the dominant community are deemed legitimate; those of the other are viewed as suspect, even as illegitimate. Thus to accept the legitimacy of the political demands of the subordinate community — such as those of non-racialism made by South African blacks and of nationhood made by Palestinians — entails (at the very least) wrenching political changes, and not the incremental inclusion of excluded groups in the fashion of 19th century Britain.

It is the problematic nature of political legitimation that distinguishes politics in Northern Ireland, South Africa and Israel. In most societies, the dominant communities would like to achieve full legitimacy, which consists of a domestic consensus that accepts the essential validity of the existing social and political orders. But the situation is different in Northern Ireland, South Africa and Israel not just because the dominant communities have had better things to do than to take the measures likely to enhance their legitimacy among the subordinate communities, and not just because they fail to establish legitimacy, but because the very prospect of extended legitimacy is experienced as a direct threat to the interests and very identities of important groups in the dominant communities. The problem is not just that resources must be expended to cultivate fuller legitimacy; the problem is full legitimacy itself. It carries unusual and vexing costs.

If the prospect of widespread legitimation is troubling to some — not to say all — interests, it remains to be explained why this is so. The answer is that some constituencies are caught in a contradiction. Israel provides a case in point.

On the one hand, genuine reconciliation with the Palestinians would imply that Israelis and Palestinians accept the legitimacy of each other's right to collective self-expression in the contested territories. On the other hand, the two sets of claims are likely to be mutually exclusive, and are difficult to discuss. The problem for Israelis is that if they were to seek confirmation of their legitimacy from the Palestinians, they would imply that the assent is valuable, or even necessary. That is why they would seek it. The corollary of this, however, is that if Palestinian consent would be meaningful when offered, then it would also be meaningful when withheld, which clearly delivers to Palestinians the right to pass judgement on the legitimacy of the disputed society.

Hence relations between Israelis and Palestinians reflect an unusual dynamic: to accept legitimation from the Palestinians is to acknowledge that they have national rights; to acknowledge these rights is to concede the right of Palestinians to challenge Israeli society; and to allow challenges is, potentially, to jeopardise the "right" of the Jewish state to exist. Better that the empowered community either deny the rights of the disempowered or claim that the disempowered can be empowered elsewhere (the Republic of Ireland, the Transkei, Jordan) than that the disempowered be put in a position that would enable them to deny the empowered — such, at least, is the import of the practice of some constituencies of the dominant community.

If the conflicts between Protestants and Catholics in Northern Ireland, whites and blacks in South Africa, and Jews and Palestinians in "Eretz" Israel begin as ones of settling and residing populations, it remains to be explained how these conflicts differ from common nationalist conflicts. The difference is that in nationalist conflicts the specific groups see themselves as more or less self-contained peoples, whose unity is forged by the belief in common descent.[3] By contrast, settlers constitute and unify themselves precisely through interaction with competing communities. Rather than arriving in the new lands with shared identities or the belief in blood ties, the settlers recognise that it was the experience of settling the new lands and of struggling with the natives that converted them from disparate and quarrelsome individuals into functioning communities linked in a peculiar and vexing relationship with the other community.

The one-time settlers who became Northern Irish Protestants, South African whites or Israeli Jews identify themselves relationally; their traditions — especially those concerning politics — tell them not just who they are, but equally who they are not. On the one hand, they develop positive identities based on things like language, common economic interests, a sense of common fate, and love of the land, none of which entails specific relations with the other populations. But, on the other hand, the early settlers founded political communities that accentuated the antagonism with the prior populations.

As they elaborated political authorities, formed political traditions, defined political membership and obligation, subsequent generations inherited the conflicts with those whom they challenged and who challenged them. Their polities became grounded in the contradiction between their loyalty and the disloyalty of the subordinate population to the established political institutions, which helped to form the political categories and identities that mediated and overcame the extensive ethnic diversity on the dominant side of the "national" divide.

Take, for example, Irish Protestants. In Ireland historically and Northern Ireland currently, almost all settlers and their descendents are Protestant; profess their allegiance to Britain; view the prospects of a united Irish republic with foreboding; respect the existing social and political orders as legitimate; and understand themselves to be the guardians of freedom and individualism against Catholic superstition and tyranny. The appearance of homogeneity was once deceptive, however.

From the 17th century until the 19th century the Protestants were divided, often bitterly, between English Anglicans and Scottish Presbyterians.[4] What broke down most of their differences and bound Protestants into one cohesive people was the emergence of Irish Catholics politically in the middle of the 19th century. Even after the salience of religious and ethnic differences lapsed, Irish Protestants fought among themselves over social and economic issues. But in spite of their enduring and intense disputes, only once — in 1798 — did they extend to the "constitutional" question, and never again was the support of Catholics solicited by the losers in disputes among Protestants. Thus not only does the contrast with Catholics reconcile divisions among Protestants, but the political unity of Protestants develops against the background of the threat posed by Catholics.

Catholic opposition is essential to the collective interests and identities of pivotal constituencies of Protestants. Not only have Catholics suffered the deprivations — politically, socially, and economically — that were the necessary corollary of Protestant privileges, but these are rooted in, and justified by, the discrepancy between Protestant "loyalty" and Catholic "disloyalty".

Protestants claimed, and historically Britain accepted, that they were entitled to land, better jobs, and preferential access to state resources because they, unlike Catholics, were loyal to Britain and to the political order it had established in Ireland. But in justifying their political power and social privileges on the grounds of their "loyalty" and Catholic "disloyalty" to the established social order, Protestants fostered an abiding contradiction. Protestant "loyalty" was meaningful only in contradistinction to the "disloyalty" of Catholics.

Thus Protestants, and especially the more marginal ones most dependent on privileges to preserve their status, fostered the traditions and interests that integrated them into a coherent community at the expense of excluding Catholics. Catholic hostility was both the outcome and the condition of Protestant solidarity. Protestants contained their differences by accentuating those with Catholics.[5]

Protestants have opposed Catholic empowerment despite high costs. From 1921 through 1969, the Unionist Party governed Northern Ireland monolithically. Although Catholics held the franchise and political rights, the unity of the Unionist Party, its capacity to manipulate the "constitutional question" to its advantage, the reluctance of Catholics to participate in official institutions, and the refusal of Protestants to share power with them in any case reduced the substantive importance of Catholic political rights. The Protestant majority might have allowed Unionists to control the Northern Ireland state — Stormont — indefinitely had it not been for the decline which beset Northern Ireland's economy in the 1950s and 1960s. To revive it, Prime Minister Terence O'Neill sought foreign investment, toned down sectarianism, and adopted more technocratic language while at the same time trying to avoid giving offence to Unionist traditions. The strategy failed.

Catholics, aroused from their habitual combination of political apathy and hostility to the Northern Irish state, initiated a campaign for equal rights. Meanwhile, conservative Protestants, fearing that their traditional privileges were slipping, demanded that harsh measures be taken against Catholic demonstrators. Soon communal violence erupted, Britain intervened, and the provincial government collapsed.

Thereafter, Britain proposed various compromises for Northern Ireland, most notably a power-sharing government in 1974 and the Anglo-Irish Accords in the late 1980s, only to see Protestants reject the deals, reinforce the cycle of communal conflict, and perpetuate Britain's reluctant governance of Northern Ireland. Protestants, it must be concluded, would rather forfeit powers themselves than run the risk of empowering Catholics.

As the unity of Northern Irish Protestants is fostered through the presence of Catholics, so is that of white South Africans. White South Africans are divided by language, religion, ethnicity, culture. Often the competition between the two largest groups of whites — Afrikaners and English-speakers — has been intense and even violent. It has been argued convincingly that Afrikaners perceived themselves to be victimised by the English-speakers, that they sought redress by mobilising as a group, and that the content of South African politics results in large measure from the ethnic mobilisation of Afrikaners.[6]

The ethnic rivalries between whites commenced in the 19th century, and were highlighted by Afrikaner resentment of British interference in the Cape,[7] the withdrawal of many Afrikaners to the interior for the purpose of founding their own republics, the machinations of the mining industries, and Afrikaner suffering during the Boer War. The ethnic rivalry between Afrikaners and English-speakers overlapped, moreover, with class conflicts, and often seemed to be creating two irreconcilable white communities. But the divisions were contained, if not resolved, by deals struck by Britain after the Boer War. Britain enfranchised whites and disenfranchised other racial groups (except in the Cape),[8] thereby using race and racism to integrate otherwise disparate whites into a coherent, identifiable and more or less united coalition.[9]

The effect on South Africa of constituting citizenship along racial lines scarcely can be exaggerated, either for black-white or intra-white relations. For blacks, it meant that they were powerless politically, even though they were essential economically. Thus when South Africa began legislating an apparatus of labour exploitation and political control, they were unable to resist successfully. For whites, it meant that they were poised to work out political agreements and to forge a composite political interest, that even poor Afrikaners were empowered over Africans, and that English-speakers were denied the black allies that they would have needed to break Afrikaner control of the state if — and when — Afrikaners united politically against the English-speakers.

But most of all, the establishment of a racially structured political order rooted what united whites among themselves and what differentiated them from blacks in one and the same thing: political enfranchisement. Enfranchisement, moreover, had a dual value. Instrumentally, it raised whites over blacks by allowing the one to claim a disproportionate amount of resources.[10] Existentially, it registered the superiority of whites. Blacks, by implication, became the implicit backdrop, the "other", by which whites developed the shared meanings and traditions that defined what it means to be "white".[11]

The point is not that political relations in South Africa — or the other two societies — are necessarily frozen and unchanging, as if they had been innoculated against historical change by an injection of settlers. But pronounced changes in the social, economic and demographic spheres accent the relative lack of political change in South Africa. At the outset, racial domination was so routine that it was taken for granted. With white supremacy secured, serious controversy was stirred less by racial conflicts than by the class and ethnic conflicts among whites. What contained these conflicts was the shared need on the part of all whites to control black labour and the state. "South African whites," wrote one scholar, "had wide divergences of interests and were, at times, united only by their agreement on the necessity to maintain the subordinate position of the blacks."[12]

The period of routine domination ended with the election in 1948, installing the first in a series of Nationalist governments. The Nationalists replaced the generally pragmatic racial policies of previous governments with the ideologies of apartheid and separate development, which together denied that blacks were citizens and thrust many into homelands despite the considerable costs of relocation. Meanwhile, the Nationalists promoted the ethnic interests of Afrikaners and succeeded in reducing the number of poor Afrikaners and in developing an Afrikaner bourgeoisie.

Over time, however, the costs of ideological apartheid weighed very heavily on the South African state and economy. Labour shortages vexed capital, widespread black poverty limited the internal market, demographic changes increased the size of the black majority, international criticism intensified and was supplemented by sanctions and divestment, and the 1976 rebellion in Soweto raised the fear that apartheid was inspiring a generation of blacks determined to overthrow capitalism along with apartheid.

In response, the Nationalists introduced a series of social and economic reforms — such as the legalisation of black trade unions, the recognition of section 10 rights, and the abolition of much of "petty apartheid" — explored less ideological, more technocratic modes of control, and signalled their willingness to "talk about talks", to negotiate on more inclusive constitutional structures, and to enlist increasing numbers of black functionaries as clients. Thus the state became less an instrument of Afrikaner ethno-nationalism and more a structure with a life of its own.

But the Nationalists, because they must respond to their electorate, also continue to balance their openings to blacks with statements about the need to preserve "group rights" and to recognise that South Africa is a nation of minorities. The resulting political dangers are obvious. The government, in espousing pragmatic and technocratic as opposed to racial values, must contend at once with a black majority demanding fundamental — if not revolutionary — transformations, and an intransigent white minority fearing that political concessions will doom them and their way of life. Thus, while the policies of the past decade bespeak an attempt to introduce and control political change, it is uncertain whether — and how — consensual changes will occur in political institutions.

The relation of Israeli Jews to Palestinians is different. Unlike Northern Irish Protestants and South African whites, Jews have existed as a recognisable and internally integrated people for thousands of years; have developed an acute sense of solidarity and common fate; and have cultivated the skills necessary to survive persistent outbursts of anti-Semitism. Almost all Jewish Israelis have gone on to accept the need for a Jewish nation-state and to affirm the state of Israel as the legitimate expression of Jewish nationalism. Nonetheless, Israeli Jews remain divided between Ashkenazi and Sephardic, with even these categories overshadowing important differences within them, which has the effect of making the Jewishness shared by Israeli Jews often seem as divisive as it is unifying.[13]

The Arab states threaten Israel: that is obvious and undeniable. Perhaps that is why Israel responds coherently and effectively to them. But the threat posed by Palestinians within is subtle and political, not blunt and military. Thus the Palestinian question raises real problems for Zionists, especially those from the labour tradition. While the

Revisionists and their heirs can acknowledge that Palestinians constitute a people, albeit one with an inferior claim to the lands of Israel, Labour Zionism cannot dispose of the Palestinian question as easily. Labour Zionism maintained that Jews constitute a nation who, like other nations, deserve a state of their own. They justified their demand for a Jewish state less on the grounds of the unique sufferings of the Jewish people than on the structure of human societies.

Since politics is organised by nation-states, and since peoples without nation-states of their own necessarily suffer deprivations, it is, according to the Labour Zionist tradition, as imperative for Jews to have a homeland of their own as it is for any other people. The claim of the Jewish people to a nation-state of their own receives added poignancy, of course, from the tragedies they endured when living at the sufferance of other people, but Jewish sufferings only enhance the claim for Labour Zionists; they do not constitute it. The Labour Zionist claim is, in principle, the same as that of any other people: the right to self-determination.[14]

Having based its demand for a Jewish state on the rights of all peoples to self-determination, Labour Zionism could not consistently deny to the Palestinian people a state of their own. The solution — since Zionists wanted the same land as Palestinians and feared that Palestinians would not recognise them — was to deny that Palestinians held the same status as Jews. Labour Zionists premised the state of Israel on the denial of Palestinian political aspirations — indeed, of Palestinian peoplehood[15] — and in the process helped to establish the terms for integrating Jews from diverse backgrounds into one society. Jews, unlike Palestinians, comprise a genuine people and a legitimate nation, one that includes people from diverse backgrounds, histories, and regions. It is difficult, therefore, for Labour Zionists to negotiate the political future with representative Palestinians without undermining their core political assumptions, although the *intifada*, the apparent collapse of the "Jordanian option" and fear of the consequences of the *de facto* annexation of the occupied territories by the Likud recently have induced elements of the Labour Party to send signals that it might consider acceding to some sort of — limited — Palestinian homeland.

It is, however, much easier for Labour Zionists — and for that matter, for the Likud — to deal with Palestinians within the pre-1967 borders. The "Israeli Arabs" can vote, form political organisations, sit in the Knesset, and participate in the political life of Israel, although in practice patronage and manipulation employed by the government as well as the logic of a Jewish state have reduced the political significance of Palestinian rights.[16] By contrast, Israeli Arabs in the occupied territories lack the status of citizenship — in accordance, it should be added, with international law as well as their own commitments.

Thus, notwithstanding important differences, comparable dialectics develop in Northern Ireland, South Africa and Israel: in each, the dominant community is constituted through political membership and, in each, the dominant community would risk its unity and control by recognising the legitimacy of the collective claims of the subordinate community.

Important economic interests of Northern Irish Protestants, South African whites and Israeli Jews originate in political power, and particularly in control over the state. Generally, state power is conceived as something that is captured by existing classes or peoples, which in turn use it to promote their established interests. The situation for these three peoples is different, however. They did not begin life in Ulster, Southern Africa or Palestine as self-conscious classes with clear economic interests, but as heterogeneous, even inchoate, collections of individuals who developed and discovered their class unities in large measure through political organisation. Political power created as well as expressed the economic interests of the dominant actors in Northern Ireland, South Africa and Palestine.

In each society political power is used, of course, for the normal purposes of protecting property rights, pursuing particular interests, and establishing order. But it also forms and protects classes comprised of those with political access and encourages deprivations to classes comprised of members from the community with restricted — or no — political access.[17] Thus these, like most, states are both instruments and structures.[18] As instruments, they promote the interests of those constituencies that capture them, for example, by enforcing labour relations to the advantage of some workers and the disadvantage of others.[19] As structures, they constitute forces in their own right, for example, by establishing classes through patronage.

But where the states in divided societies differ from most of their counterparts is that, as instruments, they respond to groups that are strategically located politically although not necessarily economically; and, as structures, they project themselves as partisans of the classes they sponsor, nurture, and protect, and encounter resistance when they try to present themselves as independent and impartial bureaucracies governing in the interests of all citizens. With states constituting as well as reproducing class interests, they erode the characteristic distinction between economic and political interests. Political access, therefore, comes to comprise an integral part of class interests: "private" economic standings are tied — inextricably, for some — to political statuses.

Political power is the source of privileges, particularly for those who lack substantial amounts of property. For this reason, the process of industrialisation not only is compatible with the privileges of the dominant community, but can extend and deepen them. Industrialisation reinforced rather than undermined the economic, cultural and political advantages that the members of the dominant community already enjoyed.[20] Nonetheless, the beneficiaries of political power worry constantly that their living standards will deteriorate. The problem is that, while some members of the superordinate community employ subordinate labour, even more members fear the possibility of competition,[21] and not just industrial workers. Bureaucrats, lower level supervisors, the subsidised middle class, farmers — members of all of these groups fear that they can be replaced by members of the subordinate community.

The protected members of the dominant community feel vulnerable because the job skills of many of them can be acquired by the subordinate population, which often has no feasible alternative to working for lower wages. What elevates many from the empowered community, including large numbers from the middle class, above the subordinate working class, peasantry, and underclass is not usually property as such; that, in fact, is why they are vulnerable: they are dispensable economically. To protect themselves from the implications of having structured the labour market against their rivals and exposed some of their own to the danger of undercutting, the threatened members of the dominant community hold firmly to the political power that shields the vulnerable from the most severe consequences of cheap labour. Thus, rather than rooting their political position in economic privileges, they root privilege in politics: politics, not property, is the ultimate source of power in these societies.

The point is particularly relevant to South Africa. Although many studies take for granted that economic forces are decisive in the politics in South Africa, the relationship between economic interests and political power is more complicated than is often supposed. The political implications of the fact that the economic positions of many white South Africans originated in, and still depend on, benefits provided by the South African state are of critical significance, as indicated by the fact that "over one-third of the economically active white population were employed in the state sector".[22]

South African history consists not only of white economic interests capturing the state and using its resources to consolidate their position, but also of political groups using the state to convert themselves into important economic actors.[23] With the economic position of significant groups of whites — including labour, bureaucrats, and employers and employees in parastatal firms — presuming state patronage, with patronage implying prior political control, and with white political control jeopardised by black enfranchisement, some whites fear that their economic position would erode, or even collapse, if their political power were to be diluted. Thus, direct political control, while dispensable to some groups, is central to others. For this reason, the white groups with the greatest dependence on state patronage — whether for subsidies, contracts, loans or actual employment — have good reason to be averse to enfranchising blacks.[24]

Politically influential, but economically vulnerable, interests in the dominant community demand that the subordinate community be excluded from effective political participation precisely because the economic elites sometimes weigh the advantages of extending some effective representation to the excluded population.[25] These elites are motivated by the fear that political exclusion fuels instability, which, in turn, harms their economic interests. In response, they entertain the hope of forging stability from instability by extending certain rights of citizenship — within appropriate limits — to the excluded community. If they reckon that political accommodations involving the effective and genuine enfranchisement of the subordinate population can secure or, better, can increase economic benefits, they probably will propose them (except in Israel, where the issue involves territorial partition, not political inclusion).

The flexibility of the elites contrasts starkly with the inflexibility of the more vulnerable members of their community, who guard undiluted political power as the linchpin of their position. Thus, even when economic interests might suggest reason for relaxing political control, the popular classes — as opposed to the propertied — generally opt for the measures necessary to entrench their hold on political power. Thus develops one of the ironies of politics in all three societies: the more democratic the relations among the dominant group, the less the probability of democracy between groups.[26]

Conflicts with the subordinate communities help to resolve the divisions within each of the dominant communities, which arise from the overlap of ethnic rivalries with disparities in class and status. In Ireland, South Africa and Israel, one ethnic group holds — or once held — economic power over the other group. In Ireland, English Anglicans controlled the land and preferential access to the state to the disadvantage of Scottish Presbyterians until the 19th century. In South Africa, the economic and political power of the English loomed over and infuriated Afrikaners. In Israel, Ashkenazi Jews dominate both the economy and the state, much to the dissatisfaction of the Sephardim.[27]

Those suffering from inferior material conditions resented the contempt of the more affluent towards their putatively "inferior" culture. They were scorned as uncouth, uncultured and even as uncivilised as the natives. Ethnic, class and cultural tensions might have combined in each case to foster irreparable divisions and to encourage one of the groups to seek political alliances with the subordinate population. But Northern Irish Protestants, South African whites and Israeli Jews contained and mediated their fissures by forging composite political identities and brokering economic deals that fostered common interests across ethnic lines. Political power thus provides the glue connecting the dominant coalition; but it also allows the enduring differences among the partners to express themselves, often over policy towards the subordinate community.

The determination of the popular classes to deny political power to the subordinate community complicates matters considerably for their more sophisticated elites. Having displaced the old social relations with new ones and having thwarted the possibility of institutional redress for the ensuing grievances, the dominant community confirms the subordinate as a constant threat to its security. But the threat affects different interests differently, with the resulting debates furnishing the stuff of routine politics and drawing the lines for dividing the political parties within the dominant populations. What separates the Democratic Unionist Party from the Official Unionist Party in Northern Ireland, the National Party from the Conservative Party in South Africa, and Labour from Likud in Israel is not, primarily, economic or social policy, as would be the case in most liberal democracies: it is policy on the "security" question.

The vigorous debates on the advantages and disadvantages of this "carrot" or that "stick" manifest the deep fissures among the politically enfranchised population. But they also reveal the interests that all share regardless of class, status or ethnicity in perpetuating a stable order protecting their general interests. The security question, in other words, both summons forth and suppresses disagreements. On the one hand, it exposes the varied vulnerabilities of different groups within the dominant community to competition from the other community. On the other hand, the fragility of the established order mediates the conflicts by accentuating that, whatever the other internal differences among them, almost all members of the superordinate community fear the spectre of revolution represented by the subordinated community.

The agreements that knit disparate ethnic and social groups into cohesive communities are revealed by the reforms that threaten to dissolve the communities into their constituent parts. In this context, it is notable that one of the most conspicuous of the similarities between Northern Ireland, South Africa and Israel is the lack of effective institutional power held by Catholics, blacks, and Palestinians. What is striking is that, while the particular conditions and the political strategies employed to secure a monopoly vary, they converge in the practice of denying their antagonists the actual capability — if not always the formal right — of pressing political demand in legitimate institutions. Moreover, some groups in each society proceed to posit the exclusion of the "other" as the essential condition of the cohesion of the dominant coalition. Attempts to include the excluded, they maintain, will divide them and leave them vulnerable to challenge.

It is here that the differences with liberal societies are stark. In liberal societies, political rights are proclaimed as universal, with all save the most necessary exceptions entitled by the right of citizenship to equality under the law and to participation in the political process. That is not to say that the promises of liberal societies always are fulfilled, as the cases of women and racial minorities prove conclusively; but these are seen by liberals as failures, as blights on the essential commitments of liberal society. In divided societies, however, the restriction of effective political rights to the dominant community is not, strictly speaking, a failure: it is, instead, the essential condition of the political system. Rights are possessed by *groups* and not just individuals, and the rights of some groups limits the rights of others.

If some members of the dominant communities root their fundamental interests in practices and policies that antagonise the subordinate community, it is only a short step for them to move from advocating policies that antagonise their enemies inadvertently to ones that do it intentionally. Demands for rigorous enforcement of even the most discriminatory laws; tenacious opposition to reforms designed to soften control; and provocative rhetoric from populist politicians — excesses frequently bemoaned by the more enlightened elites as the irrational prejudices of the uneducated masses — actually represent an astute manipulation of the dialectics of divided societies.

The reason for this is that the more the subordinate community is estranged from the established order, the more valuable becomes the otherwise unmarketable allegiance of marginal members of the dominant communities. The advantage of this to these groups is that, as they provoke increasing hostility among their rivals, their leverage over the state improves dramatically. The elites, now needing allies in the face of challenges to the social order itself and not merely to the role of the subordinate community within it, must entertain the price set by their popular classes.

Having thereby confirmed and reproduced the hostility of members of the subordinate community, some in the dominant community proceed to use the existing threat to justify their privileges, pointing out that they, unlike their rivals, have earned special protection. However, this argument — and it is one made in all three societies — exposes the contradictions inherent in the position of the dominant community. For the security of their social and political positions depends, paradoxically, on the insecurity of the security situation: to resolve the security threat politically is to reduce the salience of their support, and thus their influence over the elites. From this dialectic follows the frequent pattern of provocation, retaliation and counter-retaliation. The dominant community, feeling intensely the insecurity of their predicament, tries to contain the danger. But they resort to repression in preference to negotiations, which hardens the hatred of the excluded community towards the established order and continues the cycle of polarisation.

It would be gratifying to conclude with a clear programme for political change in Northern Ireland, South Africa and Israel but the logic of the argument does not lead to one. That does not mean that change is impossible, although the logic of political forces in the three societies does seem to be moving against consensual democratic political changes. Some interests in the dominant communities can accept political reform, of course. But the economic influence of these groups often exceeds their political influence, which means that they might contribute to promoting reform but they cannot *effect* it themselves. The point is not that reforms cannot occur; only that the success of political reforms is likely to be determined by the strength of the political movement from the subordinate community along with the international support they can recruit, and not by the liberalism of the dominant elites.

Unfortunately, even the intersection of challenges from below with widening divisions in the community above cannot guarantee democratic political change. To the contrary, as the experience of Northern Ireland over the past twenty years warns, political reform is not only difficult but dangerous. The "troubles", it might be remembered, grew out of a genuine and serious attempt by Catholics to claim equal rights within Northern Ireland. It was, oddly, the very moderation of the demands by the "civil rights" movement that precipitated communal violence. The same fate, or even a worse one, might await reformist efforts in Israel and South Africa, for these societies not only face severe obstacles to change, but also lack a power such as Britain to contain the damage that can arise from failed reforms.

The choice between political reform and the prospect of deepening communal violence must be made in Northern Ireland, South Africa and Israel. In Northern Ireland, the forces aligned against change so far have proved substantially more powerful than those supporting it. The two communities are locked in conflict, with the extremes in both communities conspiring to prevent political accommodations. Although Britain has considerable resources at its disposal and although the rewards of a political peace are obvious to most political actors in Northern Ireland, the structure of political interests and identities nonetheless have blocked political compromises, such as power-sharing, for a generation.

In South Africa, the combination of powerful political movements among blacks and broadening political differences among whites creates an array of possibilities, some potentially democratic, others potentially chaotic. It might be hoped that the divisions among whites will soften their positions, but there is as much cause for fear that white interests, traditions and fears will induce them to cling to political power as to their very way of life.

In Israel, the opportunity for partition, not internal reform, is genuine. The annexation of the West Bank and the Gaza Strip would spell the end of a democratic Jewish state, would link the fate of Jews to the politics of Palestinians, would collide with the logic of Zionism, and would reject the lessons that Zionists have drawn from the experiences of Jews living with other people. Disengagement, on the other hand, is feasible psychologically and materially. Jews do not need Palestinians to be Jews; they do not depend on irreplaceable Palestinian labour or on Palestinian unemployment; and they could import the labour that they do need from an adjacent Palestinian state. This means that, unlike Protestants and Catholics in Northern Ireland or whites and blacks in

South Africa, Israeli Jews and Palestinians are, potentially, separable.

The possibility of a Palestinian state remains merely a possibility, however. The Likud is committed to holding the West Bank and to manipulating the ensuing conflicts to its advantage, and it is creating facts in Israeli politics as well as in the occupied territories that are raising the domestic political costs and narrowing the opportunities of disengagement. Disengagement is psychologically possible now; it might not be possible in a generation as Israelis become more attached to domination and as the influence of the groups that presume and reproduce conflict with the Palestinians increases. The import of their action is to convert the Palestinian question from an external one, where an accommodation is conceivable, into an internal one, where the two nationalisms will clash constantly and violently.

The alternative to annihilation and endless numbing violence is the recognition by each of the actors of the legitimacy of the existences of the others. The obstacles to recognition are clear, but so are the costs of failure. If the two sides cannot recognise each other, they cannot negotiate; and if they cannot negotiate, they will doom themselves to endless rounds of violence, destroying, in the process, not just each other but what is best in their own traditions. Unfortunately, the forces arrayed in each society against mutual recognition are at least as strong as those demanding it. But fortunately recognition does have one very powerful inducement: the alternatives are worse. Much worse.

ENDNOTES

1. I am following the convention of using the term "black" to refer to those groups not classified "white" by the South African government.
2. Albert Memmi: *The Colonizer and the Colonized,* Boston, Beacon Press, 1970.
3. Walker Connor: *The National Question in Marxist-Leninist Theory and Strategy,* Princeton, Princeton University Press, 1984, pxiv.
4. Steve Bruce: *God Save Ulster: The Religion and Politics of Paisleyism,* Oxford, Oxford University Press, 1986, p5.
5. Michael MacDonald: *Children of Wrath: Political Violence in Northern Ireland,* Oxford, Polity Press/Basil Blackwell, 1986.
6. Heribert Adam and Hermann Giliomee: *Ethnic Power Mobilised: Can South Africa Change?,* New Haven, Yale University Press, 1979.
7. George Fredrickson: *White Supremacy: A Comparative Study in American and South African History,* New York, Oxford University Press, 1982, pp162-79.
8. For a fuller treatment of this critical period, see Leonard Thompson: *The Unification of South Africa, 1902-1910,* Oxford University Press, 1960, especially pp5-17. A similar point is stressed by David Yudelman: "The position of large numbers of black labourers in South African society and their elimination as political actors or even as a contested political issue among whites is absolutely vital. The entire relationship of the state, capital, and organised white labour was wholly based on the premise that blacks were not a politically contested issue." *The*

Emergence of Modern South Africa: State, Capital, and the Incorporation of Organised Labour on the South African Gold Fields, 1902-1939, Westport, Connecticut, Greenwood Press, 1983, p19.

9. The word "race" is used here in its current sense of being linked with colour rather than the former sense in which Afrikaners and English regarded themselves as members of different races. To get at the meaning of the latter sense of the term, the word "ethnicity" is used.

10. This point was understood clearly by the white working class, and especially the mine workers. Having struggled and failed to gain a "civilised labour" policy in the 1920s, they subsequently won through political means what they had lost through economic competition.

11. For an important discussion of the centrality of politics and state control in forming white cohesion, see Sam Nolutshungu: *Changing South Africa: Political Considerations*, New York, Homes & Meier Publishers, 1982.

12. David Yudelman, op. cit. p34.

13. For examples of the ethnic slurs that the Ashkenazi and Sephardim directed against each other, see the description of the 1981 parliamentary election in Israel. Howard M Sachar: *A History of Israel Volume II*, New York, Oxford University Press, 1987, pp129-30.

14. Shlomo Avineri: *The Making of Modern Zionism*, New York, Basic Books, 1981, especially pp3-13.

15. The point about the denial of nationhood of Palestinians is drawn from a number of sources, including Yosef Gorney: *Zionism and the Arabs, 1882-1948*, Oxford, Clarendon Press, 1987, pp67, 132-43, and 213-18. Shabtai Teveth: *Ben-Gurion and the Palestinian Arabs: From Peace to War*, Oxford, Oxford University Press, 1985, suggests that privately Ben-Gurion recognised that the Palestinian Arabs constituted a nation in their own right, but refrained from acknowledging this lest it compromise the Jewish claim to Palestine at a time when Jews needed refuge from Hitler (pp170-1). Edward W Said: *The Question of Palestine*, New York, Vintage Books, 1980, stresses that the denial of Palestinian nationhood is integral to Zionism. Much of Said's argument is persuasive, but he seems to downplay some of the implications of denial, specifically, that denial implies an engagement with — rather than unawareness of — those who are being denied.

16. Ian Lustick argues that the failure of Israel's Arab minority to organise itself is "due to the presence of a highly effective system of control which, since 1948, has operated over Israeli Arabs" (p25). "The dominant attitude of the Jewish leadership was that the Arabs living in Israel were but an extension of the Arab world as a whole — an intrusion by the enemy into Israeli territory" (p54). "Israeli policy toward the Arab minority was determined by an overriding objective — to control the Arab community in Israel rather than to eliminate, integrate, absorb, or develop it" (p64). Ian Lustick: *Arabs in the Jewish State*, Austin, University of Texas Press, 1980.

17. An impressive portrait of how political power — extra-institutional as well as institutional — can be exerted to crystalise class interests is provided by Charles van Onselen: *Studies in the Social and Economic History of the Witwatersrand 1886-1914, Volume 2: New Nineveh*, especially Chapter 3, "The Main Reef Road into the working class: proletarianisation, unemployment and class consciousness amongst Johannesburg's Afrikaner poor, 1890-1914", pp111-70, Harlow, Essex, Longman, 1982. The status of Palestinians from the occupied territories working inside the Green Line is documented by Meron Benvenisti: "The West Bank Data Base Project, *1986 Report: Demographic, Economic, Legal, Social and Political Developments in the West Bank*", *The Jerusalem Post*, Jerusalem, 1986, pp11-14.

18. There is a full and vital literature on the nature of the South African state, with the highlights being Stanley Greenberg: *Race and State in Capitalist Development*, New Haven, Yale University Press, 1980; Stanley Greenberg: *Legitimating the Illegitimate*,

Berkeley, University of California Press, 1987 and David Yudelman, op. cit. It is also worth noting the work about Northern Ireland on this subject by Paul Bew, Peter Gibbon, and Henry Patterson, especially *The State in Northern Ireland: Political Forces and Social Classes,* Manchester University Press, 1979.

19. Stanley Greenberg: *State and Race in Capitalist Development,* op. cit., p274.
20. Stanley Greenberg, op cit., especially pp5-28.
21. This is an especially pertinent point for South Africa. It has been discussed by Stanley Greenberg: *Race and State in Capitalist Development,* New Haven and London, Yale University Press, 1980; and Merle Lipton: *Capitalism and Apartheid,* Totowa, New Jersey, Rowman and Allanheld, 1985.
22. Shula Marks and Stanley Trapido: "South Africa since 1976: An Historical Perspective" in Shaun Johnson (ed): *South Africa: No Turning Back,* Bloomington, Indiana, Indiana University Press, 1989, p30.
23. Adam and Giliomee, op. cit., pp 175-6.
24. This point is developed by Craig R Charney: "Towards Rupture or Stasis? An Analysis of the 1981 South African General Election", *African Affairs,* Vol 81, No 325, October 1982, pp527-45, especially pp533-5; and Craig Charney: "Class Conflict and the National Party Split", *Journal of Southern African Studies,* Vol 10, No 2, April 1984, pp269-82. Ronald T Libby: "Transnational Corporations and the National Bourgeoisie: Regional Expansion and Party Realignmment in South Africa" in Irving Leonard Markovitz: *Studies in Power and Class in Africa,* New York, Oxford University Press, 1987, pp291-307.
25. The best examples of the willingness to entertain the advantages of politically incorporating select members of the subordinate community are provided by Terence O'Neill in Northern Ireland and big capital in South Africa.
26. This formulation was suggested to me by Hermann Giliomee.
27. The dissatisfaction of the Sephardic Jews in Israel has been well documented by social scientists, but one of the most vivid portrayals of it was a description of a conversation the novelist Amos Oz conducted with the residents of a Sephardic housing estate. Amos Oz: *In the Land of Israel,* San Diego, Harcourt Brace Javanovich Publishers, 1983, pp34-6.

3 Nationalisms compared: ANC, IRA and PLO

BENYAMIN NEUBERGER

"Anyone who thinks he knows the answer . . . is ill-informed."
— A politican in Northern Ireland[1]

Concepts of nationhood and national identity

This chapter deals with the three major nationalist political-military organisations of the black South Africans, Catholic Irish and Palestinians who struggle by political and violent means to achieve their nationalist goals. While we also refer to other groupings and organisations like the Pan-Africanist Congress (PAC) in South Africa, the *Hamas* in the West Bank and the Gaza Strip and the Irish National Liberation Army, the focus is on the African National Congress (ANC), Irish Republican Army (IRA) and Palestine Liberation Organisation (PLO) as the dominant political military organisations.

The three nationalist organisations and the nationalism they represent are engaged in bitter and prolonged conflicts with three rival ideologies and movements — Afrikaner nationalism in South Africa, Zionism in Israel and Unionism in Ulster. African nationalism in South Africa, Republicanism in Northern Ireland and Palestinian nationalism are similar in that they fiercely oppose the established governments in South Africa, Northern Ireland and Israel. All three nationalisms are also connected to wider African, Irish and Arab nationalisms in their respective regions.

A clear definition of the concepts of "nation" and "national identity" in each case is essential in any analysis of these nationalist struggles.

The triumph of Irish nationalism in the South and the foundation of the Irish Republic have largely resolved the question of the national identity of the Irish Catholics in both the Irish Republic and in Northern Ireland. The Irish Catholics of Ulster consider themselves to be part and parcel of the Irish nation. The competing British-Loyalist identity has no meaningful support among the Catholic population of Ulster. The same applies to a narrow Ulsterist identity which has many adherents in the Protestant community but almost none among the Catholics.

The national identity underlying African nationalism in South Africa is much more complex. At least four versions of nationalism are competing for the loyalty of South Africa's African population.[2] One is South Africanism, which not only aims at "full rights for all South Africans"[3] but also perceives itself as representing one multiracial and multi-cultural South African nation. A different kind of nationalism is Africanism or black South Africanism, which stresses the common supra-ethnic nationhood of all black South Africans (Africans) and the existence of one "African people of South Africa".[4] A third type is Pan-Africanism, which believes in the unity of all Africans throughout the continent. "Africa for the Africans" and "from Cape to Cairo" have been the slogans of the Pan-Africanists.[5] Both South Africanism and Africanism have been well-represented in the ANC throughout its history, while the Pan-Africanist Congress and the Black Consciousness Movement have fluctuated between Africanism and Pan-Africanism. A fourth kind of nationalism would be ethnic-tribal nationalism (for example, Zulu or Xhosa), which many would regard as non-nationalist or anti-nationalist because it is organised from above by the South African government within a divide and rule strategy, and thus does not represent an authentic nationalist opposition against white minority rule.[6] While South Africanism is a pure type of territorial nationalism and tribal nationalism is a kind of ethno-nationalism, Black South Africanism and Pan-Africanism represent mixed types of ethnic-territorial nationalism.

The history of Palestinian nationalism also demonstrates that national identity of the Palestinians is a key question.[7] Until World War I, the prime identity of the Arabs in Ottoman Palestine was either parochial, or Ottoman and Muslim. The first nationalist stirrings referred to Greater Syria and the All Palestine Conference of 1919 indeed demanded the inclusion of Palestine in an independent Greater Syria. To this day Syria tries to keep alive the distant option of a Greater Syria by nurturing pro-Syrian Palestinian organisations (and by denying official recognition of Lebanese independence). Zuhayr Muhsin, the leader of the pro-Syrian Sa'iqa (one of the components of the PLO in the 1960s) at the time bluntly declared: "There is no difference between Jordanians, Palestinians, Syrians and Lebanese. We all constitute one people. We speak of Palestinian identity only for political reasons."[8]

A much stronger current in Palestinian nationalism has been Pan-Arabism. During the British Mandate, Jews and Arabs saw the struggle as Jewish-Arab, not as Jewish-Palestinian. The leader of the Palestinian Arabs at that time was Haj Amin al-Hussayni, who was as much a Pan-Arab nationalist (or even a Pan-Islamist) as a leader of the Palestinians. As an Arab nationalist, Hussayni was involved in the Rashid Ali al-Khilani revolt in Iraq and in the co-ordination of pro-Nazi activities in the whole Arab and Muslim world from his headquarters in Berlin during World War II. During the Mandate the major nationalist organisations defined themselves as Arab, not as Palestinian or Palestinian Arab (for example, Arab Higher Committee, Arab Independence Party and so on). In the 1950s, after the foundation of the State of Israel, almost everyone spoke about the Arab-Israeli conflict or the Arab refugee problem; hardly anyone mentioned the Palestinian dimensions of the conflict. Even the 1964 version of the National Covenant, the quasi-constitution of the PLO, is at least in part a Pan-Arab document.

The 1964 Covenant sees Palestine as part of the "Greater Arab Homeland" and commits itself to the goals of Arab unity ("Arab unity will lead to the liberation of Palestine and the liberation of Palestine will lead to Arab unity").[9] The PLO was created by the Conference of the Arab Heads of State with the declared aim of liberating Palestine, but not necessarily of establishing an independent Palestinian State. The PLO was also joined by committed Pan-Arabists who came from the Nasserite Arab Nationalist Movement (which later became the Popular Front for the Liberation of Palestine) and the Pan-Arab Ba'ath party (among the Ba'athists were Abu Nidal and Faruq Khaddumi who joined Fatah). Mishal stresses that in the 1960s "no Palestinian could afford to be accused by fellow Arabs of preferring parochial Palestinian interests (*iqlimiyya*) over broad Arab-nationalist ones".[10] Some member organisations of the PLO (for example the Arab Liberation Front) even refused to call themselves Palestinian. Even in the 1980s the extremist rejectionist Palestinian organisations (for example Jibril's, Abu Nidal's and Abu Musa's movements) regarded themselves as Pan-Arabist.

Since the *intifada* and the rise of the Muslim fundamentalist *Hamas*, a resurgence of Muslim identity, inclusive of Muslim non-Palestinians but exclusive of Christian Palestinians has become another nationalist option. *Hamas* talks about the "Muslim nation", which struggles against the Jews and their supporters — the "lands of the cross" (the

Christian West) and the "lands of the infidels" (the communist East).[11]

It has been unclear for years whether Palestine includes or excludes Jordan in the eyes of the Palestinian nationalists. In the 1980s the PLO stressed the separateness of Jordan and Palestine (maybe in order to counter the claims of the Israeli right that there is no justification for another partition of Palestine since Jordan is a Palestinian state), but in the 1970s it clearly emphasised that "Jordan is linked to Palestine by a national relationship and a national unity forged by history and culture from the earliest time".[12]

We have demonstrated that while the national identity of the Ulster Catholics has been relatively homogeneous and stable in the last decades, the same is not true for black South Africans or Palestinians. A dominant nationalism largely represented by the ANC and the PLO (although there are minorities holding different views) has nevertheless emerged in both these cases.

In South Africa most African nationalists regard all blacks as one African nation. The goal of the ANC at its foundation in 1912 was indeed to instil a feeling of common supra-tribal nationhood among all Africans. Membership of the ANC was confined to those who belonged to the "aboriginal races of Africa".[13] Their concept of one nation was supra-ethnic, definitely not supra-racial, and for more than fifty years non-Africans could not become members of the ANC for the simple reason that nobody regarded them as African nationals. The black-green-gold flag of the ANC clearly associates the organisation with black people (green symbolises the fertile land and gold the mineral wealth of South Africa).[14] To this day the pronouncements and publications of the ANC define Africans as a nation. ANC documents indeed refer to the Africans as a "majority nation" and an "oppressed nation" while the struggle waged by the ANC is defined as "national". The ANC perceives itself as a "national movement amalgamating various ethnic groups and aimed at curbing national subjugation and regaining political and economic independence".[15] The fact that all the celebrated historical "national" heroes are Africans indicates that South Africa's Africans are perceived as a nation and not as part of a multiracial nation. This does not mean that African nationalism cannot be liberal and supportive of co-existence with other national groups within one plural and democratic South Africa.

With regard to the Palestinians the dominant identity in the 1980s was definitely Palestinian. Pan-Arabism, the main competitor of Palestinianism, was severely weakened by the failure of pan-Arab unification schemes in the 1960s, the crushing Arab defeat in the Six

Day War, the death of Nasser and the rapid decline of Nasserism, the massacres of Palestinians by the Jordanian Army (during the Black September of 1970) and the Syrian Army (during the Lebanese Civil War) and the feeling of betrayal by Egypt when it signed the Camp David Accords and the Peace Treaty with Israel.

The rise of Palestinianism and the gradual decline of Pan-Arabism within the PLO became apparent in the late 1960s. The downfall of Shuqairy (PLO chairman from 1964 to 1967) and Arafat's rise to power was very much the result of a revolt by radical young Palestinians against the old guard, viewed as willing tools of the Arab states. The 1968 version of the National Covenant is obviously Palestinian, since it stresses the existence of a Palestinian Arab people and the autonomy of Palestinian decision-making. The Fatah leaders, who became dominant in the PLO in the late 1960s were "Palestinian firsters",[16] as opposed to the Nasserites of the PFLP, for whom Arab unity was at least as important as the liberation of Palestine. The *intifada* leaflets clearly demonstrate the constant and rapid Palestinisation of the PLO. The leaflets hardly ever refer to the Arab states and Arab unity and constantly stress the "Palestinian people", "Palestinian identity", "Palestinian greatness", "Palestinian stones", the "Palestinian flag" and, above all, the "Palestinian state".[17]

In all three nationalisms territory is of prime importance in the perception of what the nation is all about. Irish republicanism is strongly based on an identification with Ireland as a natural geographic, territorial and historical unit. O'Halloran talks about a "nationalist map image" which consists of the "whole island as a distinct geographical entity bounded by the sea and with no internal divisions".[18] The linguistic and cultural Anglicisation of Ireland made the emphasis on the geographic separation of Ireland and England all the more important for the preservation of an Irish identity.

Africans in South Africa, as in most other African countries, belong to a wide variety of ethnic, linguistic and cultural groups. For this very reason a supra-ethnic African nationalism in South Africa has to be based on the black Africans' historical experience of oppression within the territorial boundaries of the South African state.

Similarly, Palestinians are also not defined by language, culture or religion but by the association with Palestine as a territory and home-land. Palestinian nationalism emerged as a result of the Jewish-Arab struggle for the control of Palestine. What made the Arab inhabitants of Palestine different from other Arabs was exactly the long territorial conflict with Zionism within Palestine/Israel, in the same way that black South Africans are different from other Africans primarily because of the black-white conflict within South Africa.

Although territory plays an important role in the national identity of the Irish, black South African and Palestinian nationalists, their nationalism cannot be defined as purely civil-territorial. In the eyes of African nationalists, black South Africans are a *Staatsnation vis á vis* the tribalists, but the emphasis on their Africanness makes them simultaneously an ethno-nation *vis á vis* non-African ethnic groups. The same is true with regard to the Palestinians, who are not a *Kulturnation* distinct from Syrian, Jordanian or Egyptian Arabs but who are definitely distinct by ethnicity, culture, language and religion from the Israeli Jews who inhabit the same territory of Israel/Palestine. In the Irish case too, territory is a vital ingredient of Irish nationalism but so is religion and a sense of Irish history. Thus in all three examples the dominant nationalisms combine elements of territorial and ethnic nationalism and perceive their nations to be both and at the same time *Staatsnation* and *Kulturnation*.

It is the very importance of territory in the perception of the nation which makes it so difficult for all three nationalisms to accept the notion of partition. In effect, common to all three nationalisms is a fierce struggle against partition in any form. The republican nationalists in Ireland, and especially the IRA, have waged a fierce struggle against partition since 1920. Nationalists insist that "the national territory is the whole of Ireland and not part of it"[19] and that nobody has the right to "mutilate" the country.[20] In South Africa the ANC has consistently opposed not only the ethnic "homelands", but also any other partitionist or even semi-partitionist federal or confederal formula. The Palestinians too have fought a bitter and bloody struggle to prevent partition and, later on, to undo partition. The 1968 National Covenant of the PLO declares Palestine to be an "indivisible territorial unit" and calls for its "total liberation". Until very recently the PLO has denied any readiness to accept a two-state formula based on partition. In the 1960s and early 1970s any notion of a Palestinian state in the West Bank and Gaza (a part of the territory) was rejected out of hand. By calling such a state "Filastinistan" the PLO even compared it indirectly to the South African bantustans.

In fact all the three nationalist organisations seem today to define national self-determination in the sense that people they represent will achieve independence and establish governments ruled by kith and kin. In South Africa the major goal of the ANC is the achievement of a black-ruled South Africa by a transfer of power to the majority nation.

The demand for majority rule is more a demand of the black majority as a collective unit and less a demand for majority rule in the Western parliamentary sense (which may mean that a minority rules, for example the Conservative Party, which is the majority party in the British House of Commons, gained 44% of the national vote in the 1987 elections, while the opposition parties received 56% of the vote).

Ali Mazrui introduced the term "pigmentational self-determination" to explain the demand to replace white minority rule with black majority rule.[21] This kind of national determinism (as opposed to democratic determination) is also typical for the IRA which aims to establish Irish Catholic rule in a United Ireland. The PLO too strives for a nation-state and its publications hardly refer to its preferred form of government. All the three nationalist organisations give more emphasis to self-rule by the collective (nation) than to the ingredients of a liberal democracy: multi-partyism, free elections, individual rights, protection of minority groups and equality before the law. The emphasis is on *national* self-determination (national determinism) and not on democratic *self-determination*.[22]

Although in all three nationalisms, nation-building is to a large extent based on geography and the historical experience within a political-territorial entity, efforts are made to endow the "political nation" with a cultural content. The efforts of Irish nationalists to revive the Gaelic language and to encourage the growth of Gaelic folklore, sports and music have aimed to make Ireland "not only free but Gaelic as well".[23] Today, African nationalism in South Africa also emphasises the role of African languages and culture in a future South Africa, and the same is true of the Palestinians who have tried in recent years to stress Palestinian folklore, literature and theatre in order to underline the case for Palestinian nationhood.[24]

History plays an important role in all nationalisms, old and new. The old Irish nationalism has its origins in the late 18th century revolt of the United Irishmen and is based on vivid memories of Cromwell's massacres, the Penal Laws and the Famine.[25] Irish nationalism sees Ireland as "united since the world began" and aims to return to a pre-colonial "Golden Age". Ulster is for Irish nationalists a "repository of the Gaelic past", the home of most High Kingships of Ireland, the place where St Patrick and other Irish Saints are buried.[26] The new African nationalism in South Africa perceives modern nationalism as a continuation of resistance against European conquest by African leaders such as Shaka, Moshoeshoe, Khama, Sobuza, Mzilikazi and Sikhukuni.

While the more encompassing (South) African nationalism looks for its roots in the tribal-ethnic opposition to European conquest, the opposite is true of Palestinian nationalism. Palestinian nationalist history has no historic depth and the only national heroes rose in the 1930s and 1940s (Sheiks Az al-Din al Qassim and Abd al-Kadar al Hussayni). The truly historical heroes of Palestinian nationalism are Muslim-Arab. They are people like Salah al-Din al Ayyubi, who defeated the Crusaders, or Ammar bin Yasir, Mohammed's "Sword of Islam" (Arafat's *nomme de guerre* is in fact Abu Ammar). The three Palestinian Liberation Army (PLA) brigades are named Hittin, Ayn Jalut and Quadissiyya — all locations of great Muslim victories.[27] The mobilisation of Muslim history is even more pronounced in the Muslim fundamentalist *Hamas*.

Religion also plays a role in the three nationalisms, although in a different way in each case. Irish nationalism is very much identified with Catholicism, and religion has been the marker which has differentiated the Gaelic natives from the English and Scottish settlers who migrated to Ireland in the early 17th century. In South Africa Christianity forges a common bond between the various black ethnic groups. Clergymen have always been prominent in the ANC (and in the 1980s in the UDF) and it is not accidental that the ANC hymn "Nkosi Sikelel' iAfrika" (God Bless Africa) invokes the Almighty. With regard to Palestinian nationalism too, religion should not be underrated. During the Mandate the Higher Muslim Council, under the Mufti Haj Amin al Hussayni, led the nationalist struggle. Although the PLO perceives itself as non-sectarian (rather than secular) and stresses that "religion is for Allah, but the homeland for all of us",[28] Fatah is basically a Muslim organisation while the secular-Marxist PFLP and DFLP attract Christians. The rise of *Hamas* in the Gaza strip and the West Bank and the Islamic Movement in Israel proper also attest to the mobilisatory strength of Islam. We have already demonstrated that Palestinian (or any Arab) history is actually Muslim history.

The national identity of black South Africans, Irish Catholics and Palestinians was forged and cemented by their bitter struggle against "foreign settlers". ANC documents stress that Africans in South Africa are "sons of the soil", the "indigenous" majority who have faced "conquest and domination by an alien people" and "foreign domination" by a "foreign minority". The "us" against "them" of any nationalism is in this case the "natives" of South Africa against the "invaders" and "conquerors". For this very reason, African nationalists view the situation in South Africa as essentially colonial. The radical PAC perceives South Africa as just another white ruled colony which has to be decolonised, while the moderate ANC concedes that South Africa constitutes a "colonialism of a special type". The ANC sees in South Africa the "classical features of colonialism including foreign domination" and a "system of discrimination and exploitation based on race". But since there is no mother country and since the "foreigners" have not only ruled but also inhabited the land for three hundred years they are "an alien body only in the historical sense". Therefore the relationship between the oppressor and the oppressed nations "is not the typical imperialist relationship".[29]

As we have already mentioned, Irish nationalists also see themselves as natives of Ireland facing British foreign settlers. Nevertheless, the nationalists are willing to accept the Protestants as fellow Irishmen and thus absorb the non-Gaelic settlers within one united Irish nation. Colonialism is invoked in order to make the point that it is illegitimate for foreign settlers to usurp and separate a part of Ireland and in particular to justify the struggle against a British presence which is regarded as colonial and imperialist.

Of the three organisations studied in this chapter, the Palestinian nationalists were the strongest in their opposition to the Jewish-Zionist and later Israeli presence in the country. They characterised the Israelis as illegitimate invaders, colonialists, settlers and conquerors, and totally rejected the Zionist claim that the Jewish immigrants were returnees to the historic homeland of the Jewish people. Until recently Palestinian nationalists have insisted on the destruction of Israel as a state and on the expulsion (or repatriation) of Israeli Jews to their countries of origin. Only in the last few years, and in a very cautious and reversible way, has the more moderate wing of Palestinian nationalism (Arafat's Fatah) somehow come to terms with the existence of Israel as a state and as a nation. The more radical Palestinian organisations (the Popular Front for the Liberation of Palestine, the PFLP General Command and the Abu Nidal and Abu Musa groupings) still perceive Israel as a colonial entity which has to be liquidated.

Nationalist goals

It is important that we also analyse whether the goals of the ANC, IRA and PLO are nationalist-collective (national rights, national liberation) or democratic-individualistic (civil rights, legal equality).

The early ANC leaders in South Africa pursued a struggle limited to civic rights, like a qualified franchise, a Bill of Rights and equal pay for equal work. In these years the ANC was African-national more in its aspiration for national unity within the African community and for African civil rights, than in waging a national struggle for liberation.

In the 1940s and 1950s, and even more so since the 1960s, the ANC perceived its struggle as a war of national liberation of an oppressed nation, not as a civil rights campaign. ANC president Oliver Tambo stated clearly that "this is not a civil rights struggle at all . . . our struggle is basically, essentially, fundamentally a national liberation struggle".[30] The ANC declared that "the present stage of the South African revolution is the national liberation of the largest and most oppressed group — the African people".[31]

Although the ANC perceives the struggle to be national and although the Freedom Charter acknowledges the existence of "national groups" in South Africa, the ANC is nevertheless opposed to the granting of national rights to collectivities. Group rights are regarded as a way of preserving white rule.[32] Heribert Adam has pointed out that there definitely is a logical inconsistency in simultaneously emphasising racial groups (by demanding the "transfer of power" from the white minority to the black majority), and at the same time denying the legitimacy of any group approach when it comes to defining the structure of the future of South Africa.[33]

In the case of the Irish Catholics in Ulster the choice between a nationalist struggle for the unification of Ireland and a campaign for civil rights within Ulster is a fundamental one. Until the early 1960s Catholics in Ulster did not make a major effort to fight for civil rights in Ulster because in their eyes the Ulster statelet was anyhow illegitimate and doomed to collapse. In the 1960s, for a short interim period, there was strong Catholic support for a civil rights campaign. The Northern Ireland Civil Rights Association, which was founded for that purpose, campaigned for "one man, one vote", equality in housing and employment, for freedom of association and the dispersal of the B-Specials.[34]

Polls at that period showed that about one third of the Ulster Catholics were ready for a settlement within a reformed and democratic Ulster granting freedom and equality for all.[35] The inter-communal Alliance Party and a part of the Social Democratic and Labour Party were ready to represent an accommodationist approach, provided that an Irish dimension was at least symbolically recognised. Hence, these parties supported the Sunningdale Agreement and participated in 1974 in Brian Faulkner's first inter-communal government in Ulster history. In the 1970s even a wing of the IRA (the Officials) was willing to

de-emphasise unification and redirect the struggle towards the achievement of an Ulster Bill of Rights.[36]

The more extreme wing of the IRA (the Provisionals) strongly
opposed the civil rights approach. They were determined "to smash
Stormont" and achieve national unification by force. In the 1960s and
early 1970s everyone in Ulster was aware that Ulster's Catholic
population was divided into nationalist areas and civil rights areas.[37]

Since the collapse of Faulkner's power-sharing government in 1974,
the rapid rise of Protestant Unionist extremism (Paisleyism), the
growth of inter-communal violence and large-scale internment of
Catholics by the British Army, the civil rights approach has rapidly lost
popular support and the nationalist group approach has become
dominant in the 1980s.

Civil rights never played a significant role in Palestinian nationalism.
The nationalist Palestinian organisations, including the PLO and all its
component organisations, pursued collectivist nationalist goals, whether
maximalist (the total liberation of Palestine or the destruction of Israel)
or minimalist (since 1967 the end of Israeli occupation of the West Bank
and Gaza and the foundation of an independent Palestinian state in that
part of Palestine). Civil rights as a goal was, and still is, stressed only by
the Israeli Communist Party and only with regard to the Arabs of Israel
(within the Green Line). The PLO has never attached any importance
to even such a limited civil rights goal.

Since what matters to the ANC, the IRA and the PLO today are
collective nationalist goals, one must ask what these goals are. It seems
that while the goals of the ANC and PLO have undergone significant
changes, the basic objectives of the IRA have remained relatively stable.

The ANC was radicalised in the 1950s. From its foundation in 1912
until the 1940s the nationalism of the ANC was minimalist, gradual and
liberal. The emphasis was on "equal rights for all civilized men" and not
on "one man, one vote", on the abolition of discriminatory legislation
and not on majority rule, on integration and not on transfer of power,
on participation in government and not on national self-determination.
Things began to change in the 1940s with the demands for universal
suffrage (1943) and "black self-determination" (1947).

In the 1960s the ANC used a completely different language demanding
"national liberation", "majority rule" and "transfer of power" and
culminating in the ANC slogan "Amandla! Awethu!" (power is ours).
The transfer of power rhetoric indicates a Jacobin concept of nationhood
and democracy in which a homogeneous black majority nation establishes
majority rule. This Jacobin concept differs from a pluralistic concept of

nationhood and democracy which stresses that blacks and whites belong to one plural nation and which accepts the legitimacy of group rights within a post-apartheid liberal-democratic South Africa. In spite of such Jacobin nationalist rhetoric the ANC thinks there is a place for whites in South Africa even after the "transfer of power". ANC calls on whites to stay in the country are a clear indication that the ANC rejects radical Africanist solutions which call for "government of the Africans, by the Africans, for the Africans"[38] and deny the equal right of whites to see South Africa as their native land.

In the Irish case radical republicans remained loyal to the goals pronounced by the Irish Republican Brotherhood (IRB) in 1858, Sinn Fein in 1898, the Easter Rebellion in 1916 and the first Dail Eireann in 1919. The goals of the IRA Provisionals are to this day "to drive the invader from the soil of Ireland"[39] and achieve Irish unity. Radical republicans regard "English imperialism as the root cause of Ireland's ills" and as a major obstacle to the establishment of a Thirty-Two County Republic. They see the goals of "breaking the link with Britain"[40] and establishing a "sovereign and indivisable" Ireland as inter-connected.

After 1922 the IRA was opposed to all three established governments — the British "imperialist" government, the Protestant Ulster government and the government of the Southern "partition state"[41] which betrayed the goals of the Easter Rebellion to achieve an all-Ireland independent republic. Radical republicans see in the whole of Ireland one national self, which has the right to national self-determination by a democratic majority in the whole of Ireland. Republicans therefore totally reject the legitimacy of self-determination by the Unionist majority in Ulster which they perceive as a mere part of a natural whole.

While the ANC has radicalised its goals and the IRA has left them unchanged, the PLO — which started out as the most extremist organisation — has in the 1970s and 1980s moderated its stand. When the PLO was founded in the 1960s, it was faithful to the Palestinian nationalist consensus, which simply wanted to do away with Israel and its people and replace it by an "Arab Palestine".[42] The PLO regarded its demands as "natural", "holy" and based on the principle of national self-determination and was totally opposed to any compromise formula "dividing the Palestinian people and its problem".[43]

In 1968 the PLO adopted the more moderate goal of the "democratic state where Christians, Jews and Muslims would enjoy the same rights and the same duties within the framework of the Arab nation's aspirations".[44] The goal of a democratic state did away with genocidal threats against Israeli Jewry, often voiced during the Shuqairy period of the PLO between 1964 and 1967, but still denied the right of self-determination to Israeli Jews. The latter were destined to become at best a tolerated *religious* minority within a state defined not only as democratic but also as Palestinian and Arab ("within the Arab nation's aspirations"). The democratic state also meant continual opposition to partition and to what was derogatorily called a Palestinian mini-state on the West Bank and Gaza.

In 1974 the PLO for the first time took a step towards partition by accepting the establishment of a "national independent and fighting *authority* in every part of Palestinian land liberated".[45] It was an authority and not a *state*, in order to emphasise that it was a temporary stage on the way to the "total liberation" of Palestine. Only after 1977 did the PLO begin to propagate the idea of an independent Palestinian state in the West Bank and Gaza. Still, it was part of a strategy of stages and still it was without any readiness to grant Israel recognition as a legitimate state. Only in 1988 did Arafat disclose his support for the UN partition resolution (which the Palestinian National Covenant declares "entirely illegal") and the two states for two peoples formula. The PLO still demands the right of return of the 1948 refugees to Israel proper,[46] in addition to the establishment of a Palestinian state in the West Bank and Gaza, but there is no denying that the cumulative changes in its policies from 1964 to the present amount to a strategic moderation of its positions.

In order to better understand the true goals of the three nationalist movements, it is worthwhile to look at the analogies used by the ANC, IRA and PLO. The early ANC leaders cited the examples of Ireland, India and Egypt in the 1920s, while Algeria, Angola and Mozambique became the models in the 1960s and 1970s (to be complemented by Zimbabwe in the 1980s). All the examples represent anti-colonial struggles, demonstrating that the ANC indeed perceives its struggle as a national liberation struggle and not as an American-type civil rights campaign.

In the PLO case, the current model most frequently quoted is Algeria — a fact which very well fits with the maximalism of the PLO (in Algeria there was liberation without partition and, in essence, a transfer of its European population to France). It will be interesting to see if a more moderate PLO position will lead to a de-emphasis of the Algerian model.

For the IRA in Ulster the model is around the corner — the decolonisation of the South. Ulster itself is regarded as an artificial entity like Hong Kong, Gibraltar or the American South which tried to secede from the Union in the same way that Ulster "seceded" from Ireland. It is interesting that the IRA also identifies with the ANC and PLO as fellow "anti-imperialist movements". The other models cited in the 1980s are Iran and Vietnam: for some ANC people Iran's Islamic Revolution is a hopeful example that a people's revolt may overcome the state's military machine[47] while Vietnam is seen by all the movements as the model for a successful guerrilla war of national liberation.

Nationalist dilemmas

Almost every modern nationalist movement is plagued by similar dilemmas such as the relationship between nationalism and socialism, the emphasis on ultimate or actual goals and, finally, the use of violence or the preference for constitutionalism. The question whether nationalism and socialism are natural rivals or whether they may amalgamate is a question which all three movements had to deal with.

In the ANC there were tensions between African communists and liberals as early as the 1930s. In the 1940s rivalry developed between the Africanist nationalists of the ANC Youth League and the communists. The Africanists (amongst them Nelson Mandela, Oliver Tambo and Walter Sisulu) accused the Marxists of disregarding the racial-national nature of oppression in South Africa by propagating a class analysis of the black-white problem. The Africanists also criticised the prominence of non-Africans in the Communist Party of South Africa and the adoption of foreign ideologies.[48] Similar arguments were levelled at the ANC communists and liberals by the Pan-Africanists, who left the ANC in 1958 to found the PAC. Disagreements within the ANC on the relevance of the class struggle and the need for a socialist revolution have since continued.[49] In the 1980s a radical Marxist group was expelled from the ANC because it criticised the capitalist-bourgeois character of ANC policy. A few years earlier another group split away from the ANC because it regarded the ANC as too class conscious and internationalist rather than black nationalist.

In the case of the ANC there definitely has been a communist connection since the late 1920s. While in the early years the connection was largely personal, an institutionalised alliance gradually developed in the 1950s (through overlapping membership of African communists in the CPSA and the ANC). In the 1940s and 1950s African communists were already members of the ANC National Executive Committee. In the 1950s non-African communists like Joe Slovo, Yusuf Dadoo and Reginald September were prominent in the Congress Alliance, which drafted the Freedom Charter. In the 1960s communists joined the military arm of the ANC (Umkhonto we Sizwe or Spear of the Nation) and were prominent in the Revolutionary Council commanding Umkhonto we Sizwe. The ANC and the communists have come a long way since the early twenties when the whites-only Communist Party supported the Rand Strike under the racist slogan "Workers of the World Fight and Unite for a White South Africa". At the time, conservative ANC chiefs rejected any connection with the "assassins of the Czar".

Although the ANC-communist alliance did not do away with the distinctions between communists and non-communists in the ANC, a certain synthesis within the ANC between Marxism and African nationalism has grown since the 1960s. The communists concede that there is *national* oppression and the need for *national* liberation and they call for a national-democratic rather than a proletarian revolution. In general they belong to the ANC moderates — they support non-racialism, a mixed economy and the social-democratic Freedom Charter.[50]

The rapprochement of the communists within the ANC towards African nationalism is reciprocated by the non-communists who increasingly talk about socialism, the need to nationalise big industry (rather than the whole economy) and the necessity to redistribute wealth.[51] ANC nationalists like Nelson Mandela did not forget that "communists were the only political group prepared to eat with us, talk with us, live with us and work with us".[52] The teaching of Marxism in ANC schools and the all-out support for Soviet foreign policy (including the invasion of Czechoslovakia and Afghanistan) attest to this synthesis. The gap between communists and nationalists within the ANC has narrowed and the Revolutionary Programme may well represent the fine balance between nationalism and Marxist internationalism.

In the IRA, disagreement between right and left has also led to bitter struggles and splits. In 1934 traditional republicans, who believed in pure nationalism, split from socialist republicans who combined nationalism and socialism. In 1970 there was another split between the Marxist Officials (or Red IRA), who at that time tended to give non-violent reform in Ulster a chance (in order to convince the Protestant working class that the common interest of all workers lies in a United Socialist Ireland), and the traditionalist Provisionals (or Green IRA) who remained loyal to the goal of Irish Unity by armed revolution. In 1974 another rift divided the Officials into left and right. The left founded the Irish National Liberation Army (INLA).

From the beginnings of Irish nationalism there has been a left and right. In the late 18th century Wolf Tone declared that if the nationalists got no support from "men of property" they would base their struggle on "that numerous and respectable class of the community, the men of no property".[53] In the late 19th century Connolly propagated a "red republican interpretation of Irish history", identifying British imperialism with capitalist economic interests.[54] In 1896 revolutionary republicans founded the Irish Socialist Republican Party (a party of the same name was founded in Ulster by the left wing of the IRA Officials in 1974). Some of the republican heroes of the Easter Rebellion and the Anglo-Irish War (Connolly, Pearse, Mellows) were leftist in their orientation. Liam Mellows identified republicans with men of no property, as opposed to the Loyalist "stake in the country" people.[55] After partition the IRA at times became truly Marxist and revolutionary. In 1930 the IRA declared it aimed not only towards "the freedom of all Ireland" but also towards the "overthrow of the capitalist-imperialist system".[56] The IRA's declared aim was to found a "Thirty-two County Workers' and Small Farmers' Republic"[57] and Coogan even placed it in the 1960s "to the left of Albania".[58]

In spite of the frequent left-right cleavages within the nationalist camp, a gradual nationalist-socialist synthesis emerged both in the IRA and the ANC. In the 1980s the leftist Officials moved to the right while the rightist Provisionals moved to the left. In essence, both (and also the INLA) talk about a united socialist Ireland, in the same way as all African nationalist groups in South Africa declare their support for some kind of socialism in a post-apartheid South Africa.

Within the PLO a cleavage between the rightist Fatah and the Marxist PFLP and Democratic Front for the Liberation of Palestine has existed since the 1960s. Fatah appeals to Palestinians of all classes to support nationalist unity and is ready to co-operate with feudal-conservative Arab regimes. According to Arafat there is "no meaning to the distinction between left and right in the struggle of liberation of the homeland". He insists that the emphasis should be on a nationalist "Palestine revolution" and not on "social revolution" since "it is nonsense to insist that we wage both revolutions together because if we do so we will lose both".[59] Contrary to the ANC and IRA, the PLO establishment rarely mentions socio-economic goals like the abolition or reform of capitalism, the redistribution of wealth and the general principle of social equality. The Fronts, on the other hand, very emphatically talk about a Marxist revolution in the Arab world and about the establishment of a socialist Palestine.[60]

The Palestinian Communist Party also plays a limited role within the PLO, but while the Fronts (especially Habbash's PFLP) combine extreme nationalism with Marxism, the position of the communists is relatively moderate on the national issue (they already supported a two-states solution in the 1960s).[61] In that sense, there are similarities between the South African and the Palestinian communists.

Another typically nationalist dilemma is how to relate actual to ultimate goals. The question is whether the adoption of short-term pragmatic goals may lead gradually to an abandonment of the long-term ideological goals (without necessarily officially renouncing them).

In the case of the IRA Provisionals, a gap between immediate and ultimate goals hardly exists. Determined "to preserve rather than to compromise their ideal"[62] the uncompromisingly romantic IRA Provisionals stress a final vision at the expense of a tangible here and now. The IRA Officials were more pragmatic, but lost almost all popular support with the rise of Protestant extremism in the 1970s.

In the ANC case, too, declarations about "total power", "people's power", "transfer of power" and "majority rule", and fierce opposition to any form of "power-sharing", "group rights" and federalism seem to emphasise only the ultimate goal. Nevertheless, there are disagreements within the ANC between militants, who want to fight until victory and the seizure of power, and the moderate ANC establishment, which wishes to pressure the Nationalist government into meaningful negotiations in order to reach a political settlement based on compromise and not on total victory.[63] The call for a multiracial national convention and the readiness, even eagerness, of the ANC to talk to representatives of the white community — businessmen, politicians, intellectuals and students — may be tactical[64] but it may also be an indication that at the negotiating table the ANC will settle for a liberal-democratic compromise. Such a compromise will have to indicate some recognition of group rights and constitutional guarantees that minority tyranny will not be followed by non-democratic majority rule à la Ulster.

The gap between immediate and ultimate goals is especially wide in the Palestinian case. The PLO National Covenant still adheres to the ultimate goal of Israel's destruction and its replacement by a Palestinian Arab state in the whole of Palestine. Some declarations by Palestinian leaders in the early 1980s are still supportive of the ultimate goals: Arafat rejected any compromise because "the homeland is not a subject for give and take" and Qaddumi rejected any possibility that the "victim will recognise his murderer". The al-Hassan brothers, who belong to the Fatah pragmatists, still declared in 1982 that "the Palestinian revolution was not born to recognize Israel" (Hani al-Hassan) and that "there is not one Palestinian who will concede Israel's right to exist" (Khalid al-Hassan).[65]

Nevertheless, since the 1970s the PLO mainstream, to which at least Arafat and the al-Hassans belong, has signalled its growing though reluctant readiness to settle for half a loaf without necessarily abandoning the vision of the National Covenant. Watersheds in this development are the beginnings of dialogues with non-Zionist Israelis (1973), the readiness to establish a national authority in the West Bank and Gaza (1974), the inclusion of Zionist Israelis in dialogues (1977), the acceptance of a state in the occupied territories (1977) and the indirect Israel-PLO negotiations which led to an informal ceasefire in Lebanon (1981).

At this stage all the PLO concessions were probably tactical within the strategy of stages (Arafat in 1981: "Our people will wave the flag *first* over Jerusalem, the West Bank and Gaza and then over Nazareth and the Negev"),[66] but an accumulation of tactical concessions may amount to a strategic change. The PLO attitude to UN Resolution 242 shows the same development. From a rejection of 242 because it grants recognition to Israel (in the late 1960s) to a non-acceptance of 242 because it does not mention the Palestinian people (in the 1970s) to a recognition of *all* UN Resolutions, including 242 (in the eary 1980s), and finally to a specific acceptance of 242 in 1988.[67] Khalid al-Hassan very well explained how tactics may become strategy by saying that ultimate goals do not change reality but reality changes ultimate goals. He bluntly stated that absolutism leads to absolute desperation and that policy is the art of the possible. He appealed to the PLO not to follow their dreams, but to act according to logic.[68]

Steinberg and Harkabi compare the gradual acceptance of the two-states formula without officially renouncing the National Covenant to the behaviour of Egypt, which concluded a peace treaty with Israel without abolishing the 1962 National Covenant which calls for Israel's destruction.[69] The Irish Republic, whose 1937 constitution states that the "national territory consists of the whole of Ireland" but whose moderate governments would settle for a compromise based on power-sharing and an Irish dimension in Ulster, acted in a similar way. The Israeli Herut Party, Israel's largest nationalist party and the major component of the Likud, also adhered to a visionary programme of a Greater Israel on *both* banks of the Jordan River although since the

1950s it no longer included the East Bank in its actual policies.

Another major bone of contention in nationalist-military organisations like the ANC, IRA and PLO has been the use or non-use of violence and the extent and kind of violence employed. The ANC has the most pacific and non-violent tradition. For over fifty years (1912–61) the ANC, led by Christian conservatives and liberals, followed Ghandi's teachings of non-violence. Until the 1940s the strategy followed by the ANC was not only non-violent but also basically constitutional. The 1919 ANC constitution described "resolutions, protests, propaganda, deputations and inquiries"[70] as legitimate means to be employed in its political struggle. From 1936 to 1943 the ANC even participated in the government's powerless Natives Representative Council. Only in 1949 did the ANC abandon the legalistic way by adopting the Programme of Action calling for non-violent passive resistance. In the early 1950s Mandela explained that "constitutional means to achieve liberation can only have a basis in reality for those who enjoy democratic and constitutional rights".[71]

The first mass action by the ANC, involving large-scale strikes, boycotts, pass burnings and marches, was the 1952 Defiance Campaign. Only after it was banned in 1960 did the ANC resort to violent struggle by founding Umkhonto we Sizwe (MK). The violence employed by MK is controlled and limited to military and governmental targets and any indiscriminate killing of civilians is rejected as contrary to ANC/MK policy. Nevertheless, disagreements in the African nationalist camp between realists, who support controlled violence as armed propaganda, and romantics, who dream about seizure of power by large-scale violence, continue.

In Irish nationalism there has always been a divide between constitutional nationalists and physical force nationalists. The violent romanticism of the IRA Provisionals has deep roots in Irish history — in the struggle of the Fenians in the 19th century, in the Easter Rebellion, the Anglo-Irish War and the Irish Civil War. The Anglo-Irish Treaty of 1922 divided the nationalist camp, not only between supporters and opponents of the treaty but also between constitutional opponents (for example De Valera) and the violent opposition of the IRA. In Ulster, too, the nationalists are divided to the present day between non-violent nationalists (for example the Nationalist Party and later the SDLP) and the guerrillas of the IRA.

The IRA rift in 1970 had much to do with violence. The Officials, who tended to abandon violence, lost almost all support when they failed to defend the Catholic ghettos (their opponents mocked them by equating IRA with "I Ran Away"). The Provisionals remained faithful to the violent terrorist tradition. Their war aimed to defend the Catholics, to inspire young Catholics to join their ranks, to demoralise the British "army of occupation", to shake British public opinion and to undermine Protestant faith in the British connection.[72] After the split, disagreement regarding the use of force again developed within both the Officials and the Provisionals. While the majority of Officials finally abandoned violence in 1974, a minority separated to found INLA which was committed to armed struggle.

Even the Provisionals who supported physical force nationalism were torn by disagreement between more political and terrorist factions. An interesting development did indeed start in the 1980s when Sinn Fein, the political arm of the Provisionals, rapidly gained electoral strength and was tempted to try the political venue. Loyal to the policy of abstentionism the elected Sinn Fein members of parliament refused to take their seats in Westminster, but in 1985 elected Sinn Fein local councillors for the first time agreed to take their seats in the municipalities. The future will tell whether this means that the Provisionals are finally going constitutional.[73] In the light of such a long tradition of violent nationalism, it seems dubious whether such a development will indeed occur in the near future.

In Palestinian nationalism there has never been a Ghandian tradition as in the ANC. The PLO was committed to "armed struggle as the only way to liberate Palestine"[74] and from its early beginnings Fatah called for a "popular war of liberation".[75] For twenty years the PLO has waged terrorist warfare against Israeli, Jewish and Western targets at home and abroad, and indiscriminate killings of civilians in supermarkets, movie theatres, airports, synagogues, schools, hotels, buses and planes have been part of that terrorist strategy.

Nevertheless, disagreements between the various Palestinian organisations with regard to the objectives and targets of violence began to appear as early as the 1970s. In 1971 the PLO mainstream abandoned armed struggle as the *only* way.[76] Political-diplomatic means (for example an international conference) became legitimate, though violence was not abandoned. The rejectionists, on the other hand, believed in violent struggle as the only way to salvation. Another disagreement concerned the targets of terrorism. Fatah reduced terrorism outside Israel and the occupied territories in the 1980s and all the major attacks and hijackings abroad have been attributed to the rejectionists, in particular to the Jibril and Abu Nidal factions. In 1988 Arafat even renounced terrorism. (This does not imply the abandonment of armed struggle for, according to the PLO, attacks on army personnel or settlers in the West Bank and Gaza and even Israel proper do not constitute terror.)

In conclusion one may safely say that both in the past and in the present the ANC has been the least violent organisation, while the other extreme would be taken by the Ulster Provisionals and the Palestinian rejectionists. The mainstream PLO has occupied a middle-of-the-road position since the mid-1980s.

The Irish Provisionals (and the Palestinian rejectionists) are the closest to being a pure terrorist organisation while the ANC and, to a large extent, also the PLO are broad nationalist movements.[77] The IRA is not the military arm of a political movement; on the contrary, Sinn Fein is the political arm of a military organisation. The ANC, on the other hand, is first of all a national movement in exile with mass support, political institutions, a bureaucracy, embassies and diplomatic relations, an educational system and powerful internal allies like the Congress of South African Trade Unions, the United Democratic Front and the Mass Democratic Movement. MK is a relatively minor and marginal part of the national movement. The same is true of the PLO, which has its terrorist core but which has also grown to be a quasi-government in exile with a wide network of embassies, institutions and affiliated organisations.

Both the ANC and PLO possess a moral hegemony among their people and most observers are in agreement that they could easily win an overwhelming majority in free elections and beat their moderate (for example Inkatha in South Africa and the pro-Jordanians among the Palestinians) or radical rivals (PAC and Azapo in South Africa, the rejectionists and *Hamas* in the case of the Palestinians). Both the ANC and PLO are widely recognised by the international community (for example the UN) as legitimate representatives of their peoples. The same is not true for the IRA which has not gained any international legitimacy and which even in the peak periods of its popularity did not enjoy majority support among the Catholic Irish of Ulster. Contrary to Inkatha or the pro-Jordanians, the SDLP — the major non-violent opponent of the IRA — has enjoyed more popular support than the IRA in elections. The IRA has been opposed by part of the Catholic community who share their nationalist goals but reject their violent methods.

ENDNOTES

1. M Macdonald: *Children of Wrath – Political Violence in Northern Ireland*, Oxford, Polity Press, 1986, p149.
2. For a slightly different classification of African nationalism in South Africa see L Kuper: "African Nationalism in South Africa" in M Wilson and L Thompson (eds): *The Oxford History of South Africa*, London, Oxford University Press, p464.
3. E Munger: *Afrikaner and African Nationalism*, London, Oxford University Press, 1967, p91.
4. T Karis and G Carter (ed): *From Protest to Challenge – A Documentary History of African Politics in South Africa 1882-1964 Vol II – Hope and Challenge 1935-1952*, Stanford, Stanford University Press, 1977, pXV; on Africanism see also R Gibson: "South Africa (Azania)" in *African Liberation Movements – Contemporary Struggles Against White Minority Rule*, London, Oxford University Press, 1972, pp44-50.
5. T Lodge: *Black Politics in South Africa Since 1945*, Johannesburg, Ravan Press, 1983, p244.
6. B Neuberger: *National Self-determination in Post-colonial Africa*, Boulder, Lynne Rienner Publishers, 1986, p16.
7. B Neuberger: "What is a Nation? — A Contribution to the Debate on the Palestinian Problem", *State, Government and International Relations* No 9, Jerusalem, The Hebrew University, 1976, pp90-108 (Hebrew).
8. S Mishal: *The PLO Under Arafat – Between Gun and Olive Branch*, New Haven, Yale University Press, 1986, p44.
9. Y Harkabi and M Steinberg: *The PLO National Covenant – The Test of Time and Practice*, Jerusalem, Government Information Office, 1987, p50 (Hebrew).
10. S Mishal, op. cit., p36.
11. S Mishal and R Aharoni: *Speaking Stones: The Words Behind the Palestinian Intifada*, Tel Aviv, Avivim Publishers, 1989, p217 (Hebrew).
12. A Gersh: *PLO – The Struggle Within – Towards an Independent Palestinian State*, London, Zed Books, 1985, p112.
13. J Simons: "The Freedom Charter — Equal Rights and Freedoms", *Sechaba*, March 1985, p104.
14. P Walshe: *The Rise of African Nationalism in South Africa*, Cape Town, Donker, 1987, p204.
15. Comrade Mzala: "Revolutionary Theory on the National Question" in M V Diepen (ed): *The National Question in South Africa*, London, Zed Books, 1988, p29.
16. H Cobban: *The Palestine Liberation Organisation – People, Power and Politics*, Cambridge, Cambridge University Press, 1984, p81.
17. S Mishal and R Aharoni *passim*.
18. C O'Halloran: *Partition and the Limits of Irish Nationalism*, Dublin, Gill and Macmillan, 1987, p1.
19. E De Valera quoted by O'Halloran p177.
20. P T Ginley quoted by O'Halloran p7.
21. A Mazrui: *Towards a Pax Africana*, London, Weidenfeld and Nicolson, 1967.
22. On these theoretical distinctions see B Neuberger: *National Self-determination in Post-colonial Africa*, Boulder, Lynne Rienner Publishers, 1986, pp11-18.
23. P Pearse quoted in J Darby: *Conflict in Northern Ireland – The Development of a Polarised Community*, Dublin, Gill and Macmillan, 1976, p181.
24. M Steinberg: "Arafat's PLO: The Concept of Self-determination in Transition", *Jerusalem Journal of International Relations* Vol 9, No 3, 1987, pp86-98.
25. R Munck: *Ireland: Nation, State and Class Struggle*, Boulder, Westview Press, 1985, p173.
26. C O'Halloran, op. cit., pp19, 21, 173, 185.
27. S Mishal, op. cit., p60.

28. S Mishal and R Aharoni, op. cit., p40.
29. ANC: "Strategy and Tactics of the South African Revolution" in J Slovo: "South Africa — No Middle Road" in B Davidson, J Slovo and R Wilkinson (eds): *Southern Africa - The New Politics of Revolution*, Harmondsworth, Penguin, 1977, pp133-4.
30. W Pomeroy: "What is the National Question in International Perspective" in M V Diepen, op. cit., p12.
31. R Fatton: "The African National Congress of South Africa — The Limitations of a Revolutionary Strategy", *Canadian Journal of African Studies* Vol 18, No 3, 1984, p593.
32. P Zulu: "The Politics of Internal Resistance and Groupings" in P Berger and B Godsell (eds): *A Future South Africa - Visions, Strategies and Realities*, Cape Town, Human and Rousseau, 1988, pp125-63.
33. H Adam: "Exile and Resistance: the ANC, the SACP and the PAC" in P Berger and B Godsell, op. cit., pp95-124.
34. L Longford and A McHardy: *Ulster*, London, Weidenfeld and Nicolson, 1981, pp110-32.
35. C Townshend: *Political Violence in Ireland - Government and Resistance Since 1848*, Oxford, Clarendon Press, 1983, p388.
36. R Munck, op. cit., p141.
37. T P Coogan: *The IRA*, London, Pall Mall, 1970, p309.
38. T Caris and G Carter (eds), op. cit., p469.
39. T P Coogan, op. cit., p266.
40. Ibid. p269.
41. R Munck, op. cit., p128.
42. "The Palestinian National Charter" in W Lacqueur and B Rubin: *The Israel-Arab Reader - A Documentary History of the Middle East Conflict*, Harmonsworth, Penguin, 1984, pp366-71.
43. M Klein: *Antagonistic Co-operation: PLO-Jordanian Dialogue 1985-1988*, Jerusalem, Hebrew University, 1988, p80 (Hebrew).
44. S Mishal, op. cit., p39.
45. H Cobban, op. cit., p49.
46. Klein, op. cit., p88.
47. T Lodge: "The African National Congress: Kabwe and After", *International Affairs Bulletin* Vol 10, No 2, 1986, pp4-13.
48. A Prior: "South African Exile Politics: A Case Study of the African National Congress and the South African Communist Party", *Journal of Contemporary African Studies*, Vol 3, No 1/2, Oct 1983–April 1984, pp181-96.
49. T Lodge: "State of Exile: The African National Congress of South Africa 1976-1986", *Third World Quarterly*, Vol 9, No 1, 1987, pp1-27. On the relations between nationalists and communists within the ANC see also T Karis: "South African Liberation: The Communist Factor", *Foreign Affairs*, Vol 65, No 2, 1986-1987, pp267-87 and M Radu: "The African National Congress: Cadres and Credo", *Problems of Communism*, Vol XXXVI, July-August 1987, pp58-75.
50. H Adam: "Exile and Resistance: The ANC, the SACP and the PAC" in P Berger and B Godsell, op. cit., pp95-124.
51. T Mbeki: "South Africa: The Historical Injustice" in D G Anglin, T M Shaw and C G Widstrand (eds): *Conflict and Change in Southern Africa*, Washington, University Press of America, 1978, pp131-50.
52. T Karis: "South African Liberation: The Communist Factor", op. cit., p274.
53. R Munck, op. cit., p114.
54. M Macdonald, op. cit., p13, 143.
55. R Munck, op. cit., p124.
56. R Munck, ibid.
57. B Bell: *The Secret Army — The IRA 1916-1979*, Dublin, Academy Press, 1979, p361.

58. T P Coogan, op. cit., p38.
59. S Mishal, op. cit., p39.
60. W Quandt: "Political and Military Dimensions of Contemporary Palestinian Nationalism" in W Quandt, F Jabber and A M Lesh (eds): *The Politics of Palestinian Nationalism*, Berkeley, University of California Press, 1973, pp52-78.
61. A Gersh, op. cit., p133.
62. M Macdonald, op. cit., p117.
63. H Adam: "Exile and Resistance: The ANC, the SACP and the PAC", op. cit., pp95-124.
64. For a sceptical analysis see M Tamarkin: "A Path to Peace", *Leadership*, Vol 5, No 4, 1986.
65. A Susser: *The PLO after the War in Lebanon – The Struggle for Survival*, Tel Aviv, Hakibbutz Hameuchad, pp 47-8 (Hebrew).
66. S Mishal, op. cit., p39.
67. M Steinberg: "Trends in Palestinian National Thought", *Policy Studies*, No 25, Jerusalem, Hebrew University, 1988, *passim*.
68. M Steinberg, ibid.
69. Y Harkabi and M Steinberg, op. cit., p15.
70. Study Commission on US Policy Toward Southern Africa, *South Africa: Time Running Out*, Berkeley, University of California Press, 1981, p170.
71. E Roux: *Time Longer Than Rope – A History of the Black Man's Struggle for Freedom in South Africa*, Madison, University of Wisconsin Press, 1972, p148.
72. A M Lee: "The Dynamics of Terrorism in Northern Ireland 1968-1980", *Social Research*, Vol 48, No 1, 1981, p133.
73. On the dilemma faced by the Provisionals see J A Hannigan: "The Armalite and the Ballot Box: Dilemmas of Strategy and Tactics in the Provisional IRA", *Social Problems*, Vol 33, No 1, 1985, pp33-40.
74. "The Palestinian National Charter" in W Lacqueur and B Rubin, op. cit., p367.
75. H Cobban, op. cit., p31.
76. S Mishal, op. cit., p16.
77. On the distinction between various underground movements see D Horowitz: "Communal Armed Organisations", *Archives Euopeennes de Sociologie* Vol 27, No 1, pp85-101.

4 Policing

JOHN BREWER

Introduction

The previous two chapters have identified the points of similarity that
exist between the ethnically divided societies of South Africa, Israel and
Northern Ireland, and in the nationalist opposition forces ranged
against each of the dominant groups. This chapter extends the
comparative focus by addressing the issue of policing. The central
thrust is that while policing in ethnically divided societies approximates
to a distinct style of its own, it is often more complex than first
imagined, primarily because of an over-simplified view of the nature of
conflict in these sorts of societies. The police do mediate and reflect
societal divisions but the forms of their intervention in ethnic-national
conflict go beyond that implied in the typical portrayal of policing in
divided societies.

Policing has an impact on the wider conflict between communal
groups, but also affects the political and ideological conflicts within the
respective communities. And as these societies increasingly try to
ameliorate the divisions through a series of reforms, the impact of the
police will perhaps be greater in these intra-communal conflicts.
Policing issues thus interact with the class, generational and regional
processes which produce the fragmentation of conflict in ethnically
divided societies. The arguments suggest that the police should be
recognised as dynamic actors in their own right, with a capacity to work
independently of the regime and to reflect more than the communal
interests of the dominant grrup.

Issues of policing have not been a major focus in studies of divided
societies, as exemplified by the dearth of material on the South African
police, and work within the burgeoning field of police studies is based
overwhelmingly on research in liberal-democratic societies. This is
despite the fact that Banton noted long ago that policing in divided

societies is of special interest.[1] There are a few national case studies of policing in the ethnically divided societies of South Africa, Israel and Northern Ireland,[2] but very few studies adopt a comparative approach.[3] Accordingly, there has been little attempt to establish whether or not policing in ethnically divided societies exhibits general features. However, Weitzer has recently identified the following as the "basic syndrome of policing in deeply divided societies": sectarian enforcement of the law, political partiality, an absence of mechanisms of accountability, responsibilities for internal security as well as ordinary crime, and the precarious consent of the subordinate communal group.[4]

This chapter will show that while these are salient features, they are not exhaustive in their description of the style of policing in ethnically divided societies. An analysis of policing in the divided societies of Israel, Northern Ireland and South Africa also reveals that the manner in which the police in these three societies reflect and mediate conflict is more complicated than appears at first sight.

Policing a divided society: The basic style

The essential features of policing in ethnically divided societies are easy to extrapolate from the cases of South Africa, Israel and Northern Ireland. Policing in the latter, however, tends to be relevant only as it existed prior to the imposition of direct rule from Westminster in 1972, for the modern Royal Ulster Constabulary (RUC) is in a process of normalisation and reform, although there are severe external constraints on the extent to which this can be achieved. Policing in ethnically divided societies is characterised by the following style:

1. Selective enforcement of the law in favour of the dominant group, reflected in both a relaxed attitude towards illegal activity by the dominant group, especially when directed against subordinate communities, and excessive attention given to the behaviour of subordinate groups.

2. Discriminatory practices which limit the exercise of the rights of subordinate communities, such as applying to them standards of behaviour which do not operate for the dominant group, and the criminalisation of various forms of activity engaged in by subordinate groups. Whether it be graffiti in Israel's occupied territories, flying the Irish tricolour in Northern Ireland (now no longer illegal), or the wearing of certain T-shirts in South Africa, there is a penetration by the police into the everyday life of subordinate groups.

3. Political partisanship in upholding and enforcing the distribution of political power by allowing unequal rights to political protest, the use of repression to inhibit the forces of political change and opposition, and direct police involvement in the political process.

4. A lack of autonomy from the political system, which is shown in many ways. Police and security officials can be co-opted into structures of decision-making at the political centre so that the police have a direct input into politics, as occurred in South Africa under the now defunct system of joint management committees, and the State Security Council. Members of the police are often allowed to join political organisations or other associations connected with the political elite or dominant group, such as the Orange Order in Northern Ireland, the Broederbond in South Africa, and various religious and settler groups in Israel. A background in policing and security can also constitute an important constituency from which the governing class is drawn, as happened in Israel with the Irgun and South Africa, to a lesser extent, with the Ossewabrandwag. The incorporation of senior police officers into elite social networks or the involvement of other police members in organisations which have a political connection, can make the police subject to political pressure by other means, as tended to occur in Northern Ireland under Stormont governments. Under the Vorster administration the prime minister used Van den Bergh, a long-time political associate and head of the security police, to investigate Vorster's right-wing opponents.

5. An absence of effective mechanisms of public accountability. The police in ethnically divided societies are not impervious to external control, but the state tends to rely on its own mechanisms of accountability, or the law courts for those few instances where legal action results. However, the state is occasionally pressured into responding to criticisms of the police by instituting mechanisms of public accountability, although they are often biased in its favour, as exemplified by the numerous commissions of inquiry into the conduct of the South African Police (SAP) which have failed to find fault with the organisation.[5]

6. Relatively unrestrained use of force. This is partly the result of extensive and broad powers to use force legally, the ready availability of lethal weaponry, and the absence of mechanisms of public accountability to check the illegal use of force. But in some cases it can also arise as a consequence of opponents being

stripped of moral and humanitarian worth. This process is reflected in the derogatory terms used to describe them — *taigs, kaffirs* or *arabushim* — and treatment which shows a lack of concern with their welfare. Those divided societies which face a serious threat from paramilitary organisations can also enervate opponents by the liberal application of the label "terrorist" in order to suggest that they are beyond the pale of civilized society. A good illustration of this is provided by the remark of Menahen Milson, a former head of civil administration on the West Bank: "If people are pro-PLO they are terrorists, anti-Semites and committed to the destruction of Israel."[6] Similar sentiments have been expressed about supporters of the African National Congress (ANC) and Sinn Fein. However, the police are not free to kill on a large scale, and international norms, to be discussed in the next chapter by Guelke, act as a constraining force on those divided societies which are sensitive to world opinion. This is especially the case for post-Stormont governments in Northern Ireland, which makes the RUC much more restrained in its use of force than the police in Israel and South Africa.

7. A dual role which arises from responsibilities for ordinary crime and internal security, although as threats to internal order have grown, security tends to dominate police activities, strategic planning and leadership. The management of this dual role can be effected by the development of specialist and independent units which are responsible for internal security (such as the border police in Israel), which effectively leaves the regular force untouched by the deleterious consequences that arise from discharging this form of police work. But this function can also be diffused throughout the regular force to become a part of basic police training, as occurs in South Africa. So prominent is internal security work to the SAP that in recent cases its most senior officer has been drawn from the security branch.

8. The polarisation of attitudes towards the police and their conduct. Subordinate groups view the police as agents of oppression or occupation and show a minimal commitment to them, while the dominant community tends to look on the police as its own and the guarantors of its position.

9. A social composition biased towards the dominant group because of an inability or unwillingness to recruit from among subordinate communities. Thus, the RUC currently has a Catholic membership of between 10 and 12 per cent, a proportion roughly similar to the

number of non-Jews in the Israel police, although most of the latter tend to be in the border police and thus in practice restricted to policing the occupied territories. The police in South Africa are unique among ethnically divided societies in drawing half of their regular police from subordinate communities, although black policemen have a reputation for brutality equal to that of whites, and the occupational culture of the force is still predominantly Afrikaner. Louw has recently claimed that 90 per cent of the white core of the SAP are Afrikaners, most of whom come from the working class or rural areas.[7] Thus, to parallel the special rabbinate unit in the Israel police and the RUC's exclusive Orange Order lodge, the SAP has its own organisation to promote Afrikaans culture within the force (*Afrikaanse Taal en Kultuurvereniging van die Suid-Afrikaanse Polisie*) and all its full-time chaplains within the chaplaincy unit minister in Afrikaner churches.

10. Chronic and endemic manpower shortage. This is a product of the failure of the supply of new recruits to keep up with the specially high demand for police officers in ethnically divided societies, because of the increased range of their activities, and the structural problems these societies have in attracting recruits. Recruitment tends to be restricted to only one section of the ethnic divide, which is sometimes not the most populous part of society, and the low status and reward, and high risk associated with policing limits recruitment among the ascendant group. Staff shortage leads to the police being over-stretched and often tactically unprepared, from which can follow an over-reliance on force as the first line of defence.

11. The diffusion of policing functions throughout the dominant group, as volunteer groups and other compatible agencies are drawn into a policing role. Israel has civil guards, there are several types of reserve force in Northern Ireland and South Africa, and functionally compatible units exist in all three societies. There are both formally constituted agencies which co-operate with the police (such as Shin Bet and the Hiba soldier-women in Israel, the Ulster Defence Regiment in Northern Ireland, and numerous private and local police forces in South Africa), and informal vigilante groups which involve themselves in the conflict, whether welcomed by the police or not (such as Gush Emunim, the Ulster Defence Association and the Afrikaner-Weerstandsbeweging [AWB]).

12. Close operational links between the police and the military. Staff shortage, compounded in Israel's case by under-funding of the police,[8] and the deterioration in internal security, requires a reliance on the military. As a matter of routine or under emergency regulations, the defence forces in Israel, South Africa and Northern Ireland are given a policing role inside the country and powers of civil arrest, which ensures close co-operation with the police in strategic planning and operational practice. Occasionally this uneasy relationship amounts to complete subservience to the army, as happens on the West Bank, which is why there is little conflict between the army and the police in Israel. In South Africa there is greater autonomy between the army and the police, and some hostility, but tensions between the police and the army in the mid-1970s in Northern Ireland seem to have been largely resolved.

The concept of divided society

The term "divided society" is used in two senses in the literature. It is employed in a descriptive sense to represent societies which are deeply divided, usually as a result of ethnic and national conflicts. But it is also used as an analytic concept to suggest that there are more than superficial features which these societies share in common. The thrust of the previous arguments is to claim that the manner, conduct and style of policing is one of the features which point to parallels between ethnically divided societies. Therefore, the concept can be seen to have some analytic utility.

However, the application of the term can lead to over-simplification. Divided societies are not all divided to the same extent. For example, some contain fairly autonomous communities with separate economies, social systems and political institutions, while in other societies where there are ethnic divisions the groups share (although unequally) the society's economic infrastructure and evince greater political, cultural and ideological fragmentation. At least two problems with the concept arise from this fact. The society can contain greater commonality than the term suggests, a point made with respect to South Africa by Adam and Moodley,[9] for the divisions need not permeate all through society but rather show themselves in specific spheres (most notably the political system) or specific geographical localities. Secondly, the conflicts within a divided society can be more complex than first imagined. For example, the boundaries of the groups between whom there is conflict need not necessarily be communal, for wider ethnic divisions can be refracted by differences of class, generation and region within each group. There can also be political and ideological fragmentation, as some members of the subordinate community collaborate in administering their own oppression. Thus, the divisions are not always fixed and immutable, for alliances can emerge which demonstrate ethno-political identity to be contextual and flexible. Therefore, a distinction needs to be drawn between divided societies like Lebanon and the West Bank, which show more rigid separation, and South Africa and Northern Ireland, where there is greater economic integration and political fragmentation.

Policing in ethnically divided societies can likewise be over-simplified, for the complex nature of the divisions in these societies affect police operations. Police forces like the RUC, SAP and Israel police have the dual role characteristic of ethnically divided societies, but vast regions of the society are so unaffected by conflict that they appear relatively normal. Policing in these areas therefore approaches the consensus mode that is associated with policing in liberal democracies.[10] Policing methods in such contexts as the white suburbs in South Africa, in Israel outside the occupied territories, or in largely middle-class residential areas in Northern Ireland can be relatively normal.[11] The main problem in these areas is that of adjustment for those members of the force who are transferred from one geographic area, with its associated style of policing, to another, where a different style is more appropriate.

One other consequence that follows from operating with a more balanced view of conflict within ethnically divided societies is that the divisions in which the police intervene are not just inter-communal. Police in Israel, South Africa and Northern Ireland increasingly face public order problems arising from divisions within the dominant group, which complicates the portrayal of policing in ethnically divided societies. Divisions between Ashkenazim and Oriental Jews (the latter being the group from which most Jewish members of the Israel police come), between ultra-orthodox and secular Jews, and settler groups and their opponents in various peace movements, have a similar effect in Israel as the political and ideological divisions which sustain militant loyalism in Northern Ireland and the Afrikaner right wing in South Africa. While many of these conflicts are a product of wider ethnic-national divisions, others are unrelated to this schism.

The political and ideological fragmentation amongst the dominant groups in Northern Ireland and South Africa has created problems for the police, some of which are now largely independent of their roots in a backlash by the privileged. Efforts by ultra-orthodox Jews to secure Jewish conformity with the life-style enshrined in the Halakha constitutes the principal outward sign of this in Israel.[12] But this point was made clear to the Israel police when Emil Grunzweig, a 33-year-old Peace Now activist, was killed in 1983 by a grenade thrown at him by a fanatical Begin supporter during a demonstration against the war in Lebanon. Police investigations of criminal rackets by Protestant paramilitaries or the AWB are manifestations of the problem in Northern Ireland and South Africa respectively.

There are three knock-on effects which follow from the fact that the conflicts in which police intervene in ethnically divided societies are not necessarily always communal. There is increased competition for the political representation of the police as different political factions within the dominant group vie with each other to speak on behalf of the police. Law and order is often an issue which politically divides these factions, and conflicts over who represents the police become embroiled in debates over the government's policies of law and order. The arguments between the Conservative and National parties in the 1987 and 1989 general elections in South Africa are but more extreme examples of those that exist between the Democratic Unionist Party (DUP) and the Official Unionist Party in Northern Ireland, and Labour and Likud in Israel (at least before the coalition).

A related development is the fragmentation of the political partisanship of the police themselves as wider political and ideological divisions within the dominant group become reflected among their members. The growth of support in the SAP for the Afrikaner right wing and the neo-fascist AWB[13] has been phenomenal. Former president P W Botha reputedly believed that two-thirds of the SAP opposed his reforms and supported the Conservative Party, a proportion endorsed by most political parties in South Africa.[14] Unease within the Israel police over police action against Gush Emunim, or within the RUC over the policing of Protestant demonstrations and marches are other examples. The widely contrasting opinions expressed about the Anglo-Irish Agreement by ordinary members of the RUC is a further illustration of fragmentation in the political partisanship of the police in ethnically divided societies, as is the involvement of some members of the South African police in right-wing vigilante groups and assassination squads.[15] However, it appears from the way that the new De Klerk administration is pursuing the allegations that the SAP operated death squads that the government is trying to bring the police under the rule of law and assert civilian control over the right-wing securocrats.

The third effect is the emergence of cross-cutting perceptions of the police. A two-way shift is discernible as some members of the dominant group begin to qualify their support for the police as a result of police action against them, and support in subordinate groups starts to increase as a result of the police appearing to act more impartially. The movement in each direction is not at the same pace, for change in dominant group attitudes towards the police is more obvious. Nor is it equal across all ethnically divided societies, for the shift is more apparent in Northern Ireland than in South Africa or Israel.[16] Protestant attitudes towards the use of supergrass trials and plastic bullets, for example, changed dramatically once these methods were used against them.

Opinion polls are also instructive. Adam and Moodley cite a survey which indicates that 26 per cent of whites in South Africa did not approve of the police or army, although it is unclear for what reason.[17] Opinion trends in Northern Ireland are more clear. In a 1985 survey, conducted before the change in police policy resulting from the Anglo-Irish Agreement, with which Catholics would tend to agree, 43 per cent of Catholics rated RUC duties as "fair" compared to 59 per cent of Protestant respondents.[18] Weitzer cites figures which reveal that in the same year 38 per cent of Catholic respondents approved an increase in the size of the force.[19] Levels of Catholic support for the police do not approach those expressed by Protestants, but there is a sizeable minority from within the community which is not alienated from the RUC. However, these surveys seem to indicate that these positive evaluations are made on the basis of the RUC's routine policing role, for Catholic opinion is more whole-heartedly against specific methods which are associated with the policing of internal security. The 1985 survey showed, for example, that 87 per cent of Catholic respondents opposed the use of plastic bullets, 75 per cent opposed an increase in police undercover operations, and 93 per cent rejected the "shoot-to-kill" policy.[20] The more prominent the RUC's paramilitary role therefore, the more qualified Catholic support will be. This appears to be the case for Arabs in the occupied territories, where there is less opposition to routine activities by the police.[21] The same applies to Israeli Arabs, for in a survey during 1980 they expressed high satisfaction with certain aspects of Israeli society.[22]

The police and conflict in ethnically divided societies

It is a truism to say that by their actions the police in an ethnically divided society can either resolve or exacerbate conflict. But given that conflict in these societies is often more complex than is implied by the concept of "divided society", so that the police intervene in further ways than is suggested by the basic style of policing in ethnically divided societies, the important issue is to explore the manner by which the police reflect and mediate conflict. Once this is more clearly understood, it is possible to clarify the impact which policing has on conflicts in ethnically divided societies and identify the lines along which the police may be reformed.

The basic style of policing in an ethnically divided society illustrates some of the ways in which the police are themselves a part of the conflict rather than neutral arbiters of it. The police reproduce conflict in such societies by their partisan methods, the narrow social composition of members, their relatively unrestrained and selective use of force, their penetration into the ordinary lives of the subordinate community in the pursuit of threats to internal security, and by derogatory stereotypes which perpetuate ethnic myths and feed into police demonology. But the fragmentation of political, cultural and ideological conflicts that has occurred in ethnically divided societies allows the police to intervene in other ways.

Policing policy has become an issue of political debate both within the respective groups and across the ethnic-national divide, and therefore policing tends to be treated by political groups as a litmus test of wider conflicts. In South Africa, for example, the conduct of the police is taken as a measure of the government's commitment to reform, and the inference which the democratic movement draws from police policy and practice is that the government is not serious about change, whereas the right-wing Conservative Party infers that the government is giving too much too quickly and is hampering the police in their pursuit of law and order. Thus, in as much as political groups in divided societies hold different opinions on policy, policing becomes a source of division and conflict in another sense, for it has an impact on political relationships and ideological differences. For example, it can constitute an important division between political fractions within the dominant group (as it does most notably in South Africa between the Conservative and National parties), and by their conduct and the loyalty of their affiliations the police can affect the balance between them: a point brought home by the conduct of the People's Liberation Army in the events in Beijing during June 1989. But policing also has an important bearing on the relationship between liberation organisations and legal opposition groups within the subordinate communities. The police intervene in this conflict in several ways, which are worth emphasising in order to illustrate how multifarious is the process by which the police reflect and mediate conflict.

Militant liberation movements, like the Irish Republican Army, the Palestinian Liberation Organisation and African National Congress, invariably portray the police as a force of oppression or occupation, and withdraw their support, subjecting the police to verbal and physical attack. The moderate, constitutional and legal groups express more conditional opposition (as is the case with the Social Democratic and Labour Party in Northern Ireland) or even open support (as shown by the township councillors in South Africa and, temporarily, by the conservative Arab mayors on the West Bank). Centrist parties which attempt to straddle the ethnic-national divide also fall somewhere along this continuum; the Alliance Party in Northern Ireland being toward the latter pole, while the Democratic Party in South Africa is more critical of the police. Policing issues can thus contribute to the fragmentation of subordinate groups and be an obstacle to political and ideological unity.

The position adopted by constitutional opposition groups from within the subordinate communities is problematic. The principle of law and order needs to be upheld, for this affords them legal status and provides them with their separate political identity. But any criticism of the police can be used against them by opponents from within the dominant group for implying an attack on the foundation of law and order, while an expression of even conditional support runs the risk of bringing criticism (and even physical attack) from radical liberation organisations for collaboration and capitulation.

One response to this dilemma is for moderate opposition groups to be obsequious toward, and uncritical of, the police (or even to urge that they be allowed to form their own) in order to signal that they are a co-optable ally of the dominant group, and to be able to rely on the police for protection from radicals within their own community. Township councillors in South Africa are a good example of this. Another response is to use criticism of the police as a means to overcome ostracism from radical opponents, as reflected in Inkatha's criticism of the SAP (but not of its own police force in KwaZulu) and the SDLP's refusal to encourage Catholics to enrol in the RUC. In order not to undermine their rationale as a constitutional opposition, this criticism is usually couched in specific terms; the focus is on how the partisan nature of police conduct is counter-productive to genuine law and order, so as not to suggest any breach in their support for this principle.

In this way, policing issues are deliberately invoked by members of the subordinate group to perpetuate conflict with the dominant group but overcome it with respect to opponents from the subordinate community.

Intimidation of constitutional opposition groups by radical liberation organisations, such as occurs in South Africa's black townships and in the *intifada* in the occupied territories, can also force an escalation in their criticism of the police: in which case policing issues perpetuate the underlying fragmentation of the subordinate group. In such instances the police become involved in protecting those members of the subordinate group who have been politically co-opted, extending to them the partisan policing which is normally reserved for members of the dominant group, which adds to the divisions and conflicts within the subordinate community. When this is combined with a withdrawal of partisan policing from those members of the dominant group who are in conflict with the ruling fraction, and who object to the political co-optation of some subordinate group members, the police can easily be accused by all sides of direct involvement in societal conflict.

But the nature of the accusations illustrate how varied are thought to be their interventions. For example, in the wake of the changes in police policy introduced by the Anglo-Irish Agreement, the RUC found itself attacked by the Protestant DUP for implementing alleged rule from Dublin, intervening in the conflict in such a way as to realise a united Ireland by the back door, and abused by Sinn Fein for supposedly trying to enhance the prestige of the moderate SDLP and conservative leaders in the Irish Republic, mainly under whose efforts the Anglo-Irish Agreement was introduced. They intervened in the conflict, therefore, by attempting to isolate Sinn Fein. The Catholic SDLP was largely silent on these changes in police practice, for it feared that it might undermine its position *vis à vis* Sinn Fein by expressing enthusiastic support for the RUC, providing a good example of how police issues can impact on politics in the subordinate community.

Regimes in the throes of reform tend to make larger demands on the citizenry than do unapologetically sectarian or racist systems. This is nowhere better demonstrated than with regard to members of the police in South Africa and Northern Ireland (Israel being intransigently against change). Wider conflicts can become represented within the membership, as some object to the changes in police policy or to wider political reforms. They can demonstrate resistance to internal police reforms, such as the introduction of effective mechanisms for public accountability, the establishment of a community relations branch to improve police-public relations, and the opening up of police recruitment to a more diverse population. And they can show resistance to, and actively fight against, wider political reforms which attempt to resolve societal conflicts, so that they provide a lobby against reform or use operational discretion to intervene in support of anti-reform groups. SAP reluctance to protect cabinet ministers against AWB demonstrations or RUC unwillingness to use plastic bullets on Protestant crowds offer cases in point.

In extreme instances a minority in the police can engage in freelance acts to perpetuate traditional conflicts, such as participation in, or giving aid to, vigilante groups, assassination squads or paramilitary organisations. The reverse side to this form of intervention occurs when radical liberation movements infiltrate the police and pass on sensitive information or act as agent provocateurs. This is alleged to have happened with some Catholic policemen in the old Royal Irish Constabulary, which functioned up to the partition of Ireland in 1922, although this is strongly denied by most surviving members of the force,[23] and cases have come to light of members of the SAP being in the ANC.[24] The potential for this increases as police recruitment from the subordinate communities expands.

One other impact worth noting is how police methods can undercut the co-optation of subordinate group politicians, leading to the perpetuation of traditional social divisions. This is a particular problem where the police appear to be out of political or managerial control and where local commanders have operational autonomy. The imposition of collective punishments by the police on the subordinate community, especially favoured by the SAP and the Israeli security forces, and their brutal treatment and indiscriminate use of force, can destroy what little legitimacy conservative and moderate opposition groups achieve among their constituency. This fate has befallen township councillors in South Africa.

One way to avoid the trap is to cultivate a more radical image (and to use criticism of police methods as a means to achieve this), much along the lines of the SDLP and Arab mayors on the West Bank. The latter provide a good example of how contempt at police methods, amongst other things, can politically radicalise formerly conservative groups and therefore threaten the process of co-optation. In the 1970s, the pro-Jordanian conservative Arab mayors in the occupied territories resigned in protest against the new phase of repression and terror, particularly in Arab schools, specifically stating that the actions of the security forces made it impossible for them to help the Israeli authorities to maintain law and order.[25] Reflecting on this, the moderate Arab newspaper

Al-Quds wrote in its editorial on 19 March 1976:

The Defence Minister wants the town mayors to take action to calm down the unrest. What has he done to help them? Has he stopped the settlement actions? Did he promise to prosecute the Jewish extremists? Did he arrest the Israeli soldiers who fired at small children? Has he put an end to the intrusion of soldiers into schools and beating up pupils? He is content with imposing the responsibility on the mayors, but this responsibility is beyond their ability and outside the sphere of their activities. The mayors were not elected in order to put down the national feelings or in order to disperse demonstrations. They are part of this people, share its feelings and pains and reflects them.

Conclusion

Policing is an important factor in the search for peace in ethnically divided societies, and it follows that once we have a clearer understanding of the manner in which the police intervene in societal conflicts in ethnically divided societies, we are better able to formulate the principles around which reform of the police can be structured. The cases of South Africa, Israel and Northern Ireland prompt the following observations about this process:[26]

1. When ethnic-national cleavages are reflected in the practice and conduct of the police, its interventions are more partisan, its role in perpetuating these cleavages is greater, and one group can more easily appropriate (mistakenly) the police as their own.

2. An absence of effective mechanisms for public accountability discourages the police from being autonomous of the political system and independent of contending social groups, thus inhibiting the emergence of a role as arbiter of societal conflicts in favour of selective law enforcement.

3. If those civilian or other agencies which have authority over the police merely reproduce societal cleavages and are not heterogeneous, the constituencies which provide feedback and input into police policy making will be insufficiently diverse.

4. When police use of force is arbitrary, it allows the police to use their discretion in a partisan and selective manner.

5. The less active the police are in monitoring the petty and ordinary aspects of everyday life among subordinate groups, the easier it is for its role in internal security to be discharged without the criticism that the police are omnipresent in society.

6. The more responsibility the police have for internal security, the greater the likelihood that their routine police role will be affected by negative evaluations, and that the security branch will come to dominate the policies and managerial style of the police. Hence the more this role is restricted to specialised units within the force (or outside) the greater is the chance of avoiding contamination of the two roles.

7. The more heterogeneous the social composition of the police and the wider the organisational and institutional affiliations of its members, the easier it is for the police to recognise and respond to the claims of competing social groups, and overcome the derogatory stereotypes which sustain police demonology.

8. The less resistant the police are to internal reform, the more open they will be to changes which attempt to improve police-public relations with subordinate groups, such as the development of mechanisms of public accountability, institutional support for an effective community relations branch, the use of impartial methods of investigation, and institutional support for multi-strata recruitment policies.

9. Where competing social groups do not use policing and police reforms as a political litmus test, it is easier for police issues to be removed from contentious political debate.

10. The more latitude the police are given, either by politicians or police management, to show resistance to wider political reforms, the deeper they will become embroiled in political debate, and the more likely it is that they will become involved in perpetuating and exacerbating conflict.

ENDNOTES

1. M Banton: "Policing a Divided Society", *The Police Journal* 47, 1974.

2. See, for example, G Besinger: "The Israel Police in Transition", *Police Studies* 4, 1981; J D Brewer: "The Police in South African Politics", in S Johnson (ed): *South Africa: No Turning Back*, London, Macmillan, 1988; P Frankel: "South Africa: the Politics of Police Control", *Comparative Politics* 12, 1979; M Hovav and M Amir: "Israel Police: History and Analysis", *Police Studies* 2, 1979; R Pockrass: "The Police Response to Terrorism: The RUC", *The Police Journal* 59, 1986; S Reiser: "The Israeli Police: Politics and Priorities", *Police Studies* 6, 1983; R Weitzer: "Policing a Divided Society: Obstacles to Normalisation in Northern Ireland", *Social Problems* 33, 1985, and "Policing Northern Ireland Today", *Political Quarterly* 58, 1987.

3. For exceptions see J D Brewer *et al*: *Police, Public Order and the State*, London, Macmillan, 1988; C Enloe: "Ethnicity and Militarisation: Factors Shaping the Roles of Police in Third World Nations", *Studies in Comparative International Development*, 11, 1976.

4. R Weitzer: "Police Liberalisation in Northern Ireland", mimeo p2.

5. In this respect, it is very important to note what the De Klerk administration will do about the evidence that members of the SAP have been operating death squads.

6. Quoted in M Maoz: *Palestinian Leadership on the West Bank*, London, Frank Cass, 1983, p142.

7. C Louw: "Die SAP en die Pyn van Oorgang", *Die Suid-Afrikaan*, December 1989, pp14-20.

8. On the under-funding of the Israeli police compared to the RUC and SAP see J D Brewer *et al: Police, Public Order and the State*, op. cit., p222.

9. H Adam and K Moodley: *South Africa Without Apartheid*, Berkeley, University of California Press, 1986, p196ff.

10. On consensus policing see D Bayley: "A World Perspective on the Role of the Police in Social Control", in R Donelan (ed): *The Maintenance of Order in Society*, Ottawa, Canadian Police College, 1982; R Kinsey *et al: Losing the Fight Against Crime*, Oxford, Blackwell, 1986.

11. With respect to the RUC see J D Brewer: *Inside the RUC: Routine Policing in a Divided Society*, Oxford, Oxford University Press, 1990. With respect to the Israeli police see P Shane: *Police and People*, St Louis, Morsby, 1980, p113.

12. See S Reiser: "Cultural and Political Influences on Police Discretion: The Case of Religion in Israel", *Police Studies* 6, 1983.

13. On right-wing support within the SAP see H Adam: "The Ultra-right in South Africa", *Optima* 35, 1987, p42; J D Brewer: "The Police in South African Politics", op. cit., pp276-9.

14. See J Spence: "The Military in South African Politics", in S Johnson (ed): *South Africa: No Turning Back*, op. cit., pp252-3.

15. On the allegations of SAP death squads see J D Brewer: "The Police in South African Politics", op. cit., pp272-4; N Haysom: *Apartheid's Private Army: The Rise of Right-Wing Vigilantes in South Africa*, London, CIIR, 1986.

16. On attitudes among white students in South Africa see J Gagiano: "Ruling Group Cohesion in South Africa: A study of Political Attitudes among White University Students", conference paper delivered to the Friedrich Naumann Foundation, Bonn, September 1989.

17. H Adam and K Moodley: *South Africa Without Apartheid*, op. cit., p109.

18. Quoted in J D Brewer *et al: Police, Public Order and the State*, op. cit., p75.

19. R Weitzer: "Police Liberalisation in Northern Ireland", op. cit., p16.

20. Ibid. This introduces a structural limit on the extent to which Catholic perceptions of the RUC can be enhanced because the security situation is unlikely to improve to an extent that the RUC can dispense with high profile paramilitary-style policing. Other impediments to normalisation of police-public relations in Northern Ireland are discussed in R Weitzer: "Policing a Divided Society: Obstacles to Normalisation in Northern Ireland", op. cit., and "Police Liberalisation in Northern Ireland", op. cit.

21. Cited in J D Brewer *et al: Police, Public Order and the State*, op. cit., p151.

22. Cited in ibid. In an earlier study of public images of the Israel police, Gurevitch and colleagues argued that ethnicity was less important than education as a determinant of evaluations of the police. However, the research was conducted in 1969 and the sample was restricted to Jews in four urban areas within Israel. See M Gurevitch *et al:* "The Image of the Police in Israel", *Law and Society Review* 5, 1971.

23. See J D Brewer: *The Royal Irish Constabulary: An Oral History*, Belfast, Institute of Irish Studies, 1990.

24. Discussed in J D Brewer: "The Police in South African Politics", op. cit., p267.

25. See M Maoz: *Palestinian Leadership on the West Bank*, op. cit.

26. In a discussion of police reform in Northern Ireland, Weitzer outlines four necessary conditions: the installation of a new regime which has a commitment to reform; an

acceptance of the legitimacy of the new regime and the granting to it of popular support; overcoming the resistance of the police to reform; and the removal of the police from frontline counter-insurgency duties to concentrate on routine policing. The latter two are incorporated in the following ten-point list. The other main difference in our approach is that Weitzer believes that police reform is conditional on fundamental political change, whereas this is not necessary for some of the possibilities that are noted below. See R Weitzer: "Police Liberalisation in Northern Ireland", op. cit., pp22-5.

5 International norms and divided societies

ADRIAN GUELKE

A common characteristic of Northern Ireland, South Africa, and Israel is the sharp division between dominant and subordinate communities, underlined, as Michael MacDonald demonstrates, by the identification of the dominant community as "settlers" and the subordinate community as "natives". This division plays a fundamental role in shaping the character of nationalist movements, as Benyamin Neuberger shows. It is also at the heart of the complex problem of partisan policing in such societies, described by John Brewer. International opinion has tended to be critical of the methods used by the dominant communities in defending their position against nationalist challenges from the "natives". Far from moderating the conflict, the impact of international opinion has initially tended to increase polarisation in these cases. However, in the long run, the possibility of international mediation offers a way forward for these societies once political impasse has been reached internally.

The role that international opinion has played in shaping the global state system since 1945 is worth underlining. Most of the new states of the international political system owe their existence at least as much to the interpretation of international norms, particularly the principle of self-determination, as they do to domestic political forces. Indeed, Robert Jackson has provocatively argued that "negative sovereignty"[1] is prevalent in much of tropical Africa. That is, the new states derive what authority they possess from the international community's recognition of their legitimacy, based not on their positive capacity to establish their rule over the territory in question, but on the transfer of title to them from the colonial powers. In the language developed by Martin Wight[2] these states possess international legitimacy notwithstanding the fact that frequently the governments that rule over them lack internal legitimacy. Zaire provides perhaps the most striking example of a debilitated political system propped up by the international community's commitment to its maintenance as a single entity. A different kind of problem, dealt with in this book, is presented by the case of polities that possess effective or, at any rate, reasonably effective governments but lack international legitimacy.

The existing political systems of Northern Ireland, South Africa and Israel all lack international legitimacy, though for different reasons. In the case of Northern Ireland it is the existence of the territory itself that is at issue. In that of South Africa it is apartheid, or, more accurately, white minority rule, that is the cause of the system's lack of international legitimacy, while in that of Israel it is the system's continuing rule over the occupied territories of the West Bank and the Gaza strip.[3] In each case, the present situation, though to different degrees, is seen as running counter to the principle of self-determination as it has come to be interpreted in the post-war world. It is important to note in this context that the ending of the colonial era in the 1960s facilitated the establishment of a wide international consensus on the interpretation of self-determination. The clearest statement of that consensus is to be found in the 1970 "Declaration of Principles of International Law concerning Friendly Relations and Co-operation among States in accordance with the Charter of the United Nations".[4]

The establishment of a sovereign and independent State, the free association or integration with an independent State or the emergence into any other political status freely determined by a people constitute modes of implementing the right of self-determination by that people.

Every State has the duty to refrain from any forcible action which deprives peoples referred to above in the elaboration of the present principle of their right to self-determination and freedom and independence. In their actions against, and resistance to, such forcible action in pursuit of the exercise of their right to self-determination, such peoples are entitled to seek and to receive support in accordance with the purposes and principles of the Charter.

The territory of a colony or other non-self-governing territory has, under the Charter, a status separate and distinct from the territory of the state administering it; and such separate and distinct status under the Charter shall exist until the people of the colony or non-self-governing territory have exercised their right of self-determination in accordance with the Charter, and particularly its purposes and principles.

Nothing in the foregoing paragraphs shall be construed as authorising or encouraging any action which would dismember or impair, totally or in part, the territorial integrity or political unity of sovereign and independent states conducting themselves in compliance with the principle of equal rights and self-determination and thus possessed of a government representing the whole people belonging to the territory without distinction as to race, creed, or colour.

Every state shall refrain from any action aimed at the partial or total disruption of the national unity and territorial integrity of any other state or country.[5]

Three points need to be strongly underlined. The first is that the international community has opted to define the "self" entitled to self-determination in terms of a territorial criterion rather than an ethnic or cultural one. A significant feature of this interpretation, which "asserts the right of the majority within the frontiers prevailing at a given moment",[6] is that it makes no provision for the rights of minorities.

The second point that needs to be underlined is the linking of adherence to the principle of self-determination to the legitimacy or otherwise of the use of force or violence. The clear implication of the second paragraph quoted above is that people deprived of their right to self-determination are entitled to receive external aid in their struggle against the guilty state. Significantly, when the UN General Assembly defined "aggression" in 1974, it made specific reference to the priority of this part of the 1970 Declaration.[7]

The third point needing emphasis is the rejection of any right of secession from an independent state. Unlike the inhabitants of a colony whose right to self-determination remains an ongoing right which can be exercised in favour of independence at any time, the citizens of an independent state are bound to that state seemingly for all time. The right of self-determination has become at the point of independence the right of self-determination once, at least as far as the boundaries of the state are concerned. One can point to one major exception to this principle: the emergence of Bangladesh in 1971. The fact that Pakistan itself was the product of partition, the geographical separation between the two wings of the country, and the fact that the people of the east wing actually constituted a majority within the whole country were factors in legitimising this single and controversial example of contested secession.[8]

However, the general hostility of the international community towards the principle of secession remains as strong as ever. Secessionist movements have quite an impressive record in terms of attracting popular support and of establishing temporary control over territory but their lack of international legitimacy has been a major obstacle to their ultimate success. For example, recognition of an independent state for the Kurds would require a revolution in the interpretation of the principle of self-determination. On this ground alone the odds are heavily stacked against such an outcome, virtually regardless of the local balance of forces.

The precise legal standing of the current interpretation of the principle of self-determination is a matter of dispute among international lawyers. According to Pomerance, for many representatives at the United Nations self-determination has become "*the* peremptory norm of international law, capable of overriding all other international legal norms and even such other possible peremptory norms as the prohibition of the threat or use of force in international relations".[9] But whether this view is justified or not, what is important from a political perspective is that the international community, acting through the United Nations, appears in practice to accord a higher priority to the principle of self-determination than to other norms of the present world order such as non-intervention and human rights. Thus states held to be violating the principle of self-determination find it of little avail to appeal to Article 2, paragraph 7 of the UN Charter which bars the UN from intervening in "matters which are essentially within the domestic jurisdiction of any state". Similarly, states already seen as illegitimate in terms of self-determination attract a disproportionate degree of opprobrium for breaches of human rights.[10]

At this point, it is necessary to examine the cases of Northern Ireland, South Africa and Israel individually to bring out more fully the political implications for each of the international community's current interpretation of self-determination. Northern Ireland, comprising 6 of the 32 counties of colonial Ireland, was established in 1920 as an autonomous political entity within the United Kingdom to accommodate Protestant opposition to Irish home rule. The province was given its own parliament and government and while Northern Ireland elected members to the House of Commons in London, they formed a barely noticed contingent on the Conservative benches, especially as, in deference to the province's institutions, the affairs of Northern Ireland were not discussed in the Westminster Parliament. Effectively, Northern Ireland was outside the British domestic political system. Indeed, removing the Irish question from British politics had been one of the principal objectives of British policy when Northern Ireland was established.

This history is reflected in Northern Ireland's current constitutional status as a conditional part of the United Kingdom, different from any other part in that British governments acknowledge its right to secede.[11] This may also be seen as the logical obverse of the constitutional guarantee first extended to the province in 1949 that Northern Ireland shall not cease to be part of the United Kingdom without the consent of a majority of the people of Northern Ireland. This makes Northern Ireland's position in this respect comparable to Britain's remaining colonies. They too remain bound to Britain only by the consent of their inhabitants. In effect, the province enjoys a status somewhat above that of a colony, but below that of, say, Scotland. From an international perspective, Northern Ireland is, as a report of the European Parliament described it, a "constitutional oddity".[12] In most other countries it would be quite unthinkable that part of the national territory should be regarded as, in principle, detachable. More importantly, Northern Ireland's current status runs counter to the model of the sovereign independent state with fixed boundaries that constitutes the primary and preferred mode for the implementation of the principle of self-determination in the 1970 Declaration. This plays a large part in the province's lack of international legitimacy.

This begs the question: how might Northern Ireland's position be brought into line with the present interpretation of the norm of self-determination? There is in fact more than one answer to this question. A case can be made out for four "solutions". They are the full integration of Northern Ireland into the United Kingdom; the reunification of the British Isles as a political unit; an independent Northern Ireland; and a united Ireland. The first and last are the most important in practice in influencing external perceptions of the problem. From a legal perspective, Northern Ireland is fully part of the United Kingdom, an independent state that is a legitimate member of the United Nations. Consequently one might argue the question of Northern Ireland's international legitimacy does not arise because it is simply a subordinate part of a state. Thus, the bar in the Charter on intervention by the United Nations in domestic matters provided the basis on which Britain successfully opposed the inclusion of the situation in Northern Ireland on the agenda of the Security Council in 1969, so the legal position has been of some value to Britain in preventing the internationalisation of the issue in a formal context.

Unlike the situations in South Africa and Israel, the conflict in Northern Ireland has never been deliberated on, in a formal international context. However, the European parliament has taken up the issue.[13] In this case, political perceptions of the nature of the conflict and the assumption that the parliament is more than simply an international body have overcome the inhibitions on discussing the domestic affairs of a member state. However, even in the case of the European parliament, the legal position has operated as a brake on its discussion of the problem. What would be required to make the integrationist perspective *politically* credible would be a declaration by the British government that Northern Ireland was permanently part of the United Kingdom.

Not merely have successive British governments declined to make such a commitment to the permanence of the union, but with the signing of the Anglo-Irish Agreement in November 1985, Britain has committed itself in an international treaty to facilitate a change in the status of Northern Ireland should its inhabitants so wish. Since the Agreement, support for the option of integration has grown in Northern Ireland. However, it is unlikely that any British government would lightly accept the open-ended commitment to Northern Ireland that integration would demand, even if the abrogation of the Anglo-Irish Agreement permitted it to be made.

The view that a united Ireland is both desirable and inevitable dominates external *political* perceptions of the Northern Ireland conflict. This is particularly evident at moments of crisis, such as the 1981 hunger strike.[14] The main tenets of what can be called the nationalist perspective are simple to state. In the first place, Northern Ireland was an artifically created entity. Its genesis was illegitimate not merely because it partitioned the island but because of the particular boundaries drawn.[15] Consequently, Northern Ireland's inclusion within the United Kingdom constitutes a denial of the right to self-determination of the majority within the island. In the second place, Britain's presence in Northern Ireland constitutes a form of colonial rule, a perception underlined by direct rule and the presence of British troops on the streets, as the Provisional IRA was quick to appreciate and exploit. One of the main strengths of the nationalist perspective in the light of the current interpretation of the norm of self-determination is that the meaning of territorial integrity in relation to an island seems unambiguous, at least in a geographical sense. The geographical image has tended to influence the political (though admittedly not the legal) interpretation of the term, however much geographers may deplore the whole concept of natural frontiers. It is also a question of political practice. Divided sovereignties on islands are rare.[16]

Not merely then does Northern Ireland lack international legitimacy, but there is a wide consensus of *political* opinion outside Northern Ireland that a united Ireland is the solution to the problem. (The emphasis placed on *political* is to underscore the dichotomy with academic analysis of the conflict, much of which would question whether a united Ireland was possible.) While there is considerable appreciation in the outside world that the Protestant majority in Northern Ireland opposes a united Ireland, the political basis of Protestant opposition, because of its poor fit with existing international norms, appears as anachronistic and illegitimate to much of world opinion as does white resistance to majority rule in South Africa. Its strength also tends to be underestimated, the further removed people are from the conflict. Admittedly, the Anglo-Irish Agreement has attracted a wide measure of international support, particularly in the developed world, and this is important as an endorsement of the policies of the British and Irish governments. However, such support cannot really be seen as a step towards accepting the legitimacy of Northern Ireland as a political entity, since it has been given for the most part in the expectation that the Agreement constitutes a stepping stone to a united Ireland.

What then are the most important consequences of Northern Ireland's lack of international legitimacy? It is best to discuss its limits first. For the most part it has not internationalised the conflict in the sense of drawing states outside the British Isles into the conflict. The principal exception in this regard is Libya, which has intermittently supplied arms to the Provisional IRA. These include some very large shipments in relation to the general level of violence in the province. Libya's involvement appears to be partly ideological and partly a function of its bad bilateral relations with the British govenment. The other topical exception is South Africa, with much publicity being given to apparently well founded allegations that Armscor has supplied arms to Loyalist paramilitary organisations through intermediaries in Israel and in the Christian militias in Lebanon.[17] The shipment of arms to Loyalists evidently formed part of Armscor's efforts to get its hands on surface-to-air missile technology belonging to the Belfast firm of Shorts.[18]

However, in general, it is not difficult to understand the absence of external intervention by states in the conflict. Very few states have anything to gain from intervention in the conflict that could possibly compensate the impact that such involvement might have on their relations with Britain and the Republic of Ireland. It seems doubtful that involvement has benefited Libya or South Africa. On the other hand, Northern Ireland's lack of international legitimacy makes it difficult for Britain to marshal international opinion against such interventions.

At a diplomatic level, Britain has come under considerable pressure from other states, particularly from the United States of America. Such pressures played an important role in the negotiations over the Anglo-Irish Agreement. There has also been some involvement in the conflict from outside the British Isles below state level. Once again much the most important source for such involvement is the United States. A significant proportion of funds for Republican paramilitaries has been raised in the United States through organisations such as Noraid, while the campaign by the Irish National Caucus that American firms investing in Northern Ireland should adhere to the MacBride principles, an employment code "modelled" on the Sullivan principles which were applied to South Africa, has both severely embarrassed the British government and had a considerable influence on the government's policy in relation to discrimination.[19]

However, it would be wrong to imagine, because there is relatively little direct external intervention in the conflict, that Northern Ireland's lack of international legitimacy does not have a significant impact on the conflict. Indeed, the most significant consequences of international questioning of Northern Ireland's existence are to be found within the province itself. In the first place, despite the intensity and solidity of Protestant opinion against a united Ireland, the expectation of a majority of people in Northern Ireland appears to be that a united Ireland will, none the less, come to pass.[20] The important political effect of this expectation is to reduce the scope for political compromise between the two communities because schemes for political accommodation tend to be viewed by both sides as transitional. A possible exception is the notion of an independent Northern Ireland, though even independence could be seen as a stage towards eventual reunification rather than an end in itself.

The political initiatives of the British government tend to be judged in a similar light. Despite the weight of government propaganda to the effect that the Anglo-Irish Agreement safeguarded the position of the majority opposed to a united Ireland, a considerable proportion of respondents in a poll of opinion in Northern Ireland on the Agreement concluded that it brought a united Ireland closer.[21] Ironically, the very success of the Agreement internationally has contributed to Unionist suspicion of it, since Unionists recognise that world opinion favours a united Ireland. In general, Unionists tend to be hostile towards international organisations and sympathetic to those resisting the demands of the world community such as South Africa and Israel, while nationalists tend to identify strongly with world opinion on such issues. Both Unionists and nationalists are apt to draw conclusions about the direction of the British government's policy towards Northern Ireland from its attitudes on these questions. Thus, the success with which the Conservative government steered Zimbabwe to independence in accordance with international norms in 1980 both raised nationalist hopes and aroused Loyalist fears.[22]

In the second place, Northern Ireland's lack of international legitimacy plays an important role in the legitimisation of political violence. It helps to promote a siege mentality among Protestants that provides a justification for the actions of Loyalist paramilitaries. At the same time, it gives external credibility to the Provisional IRA's claim that it is engaged in an anti-colonial struggle against British imperialism. In the third place, Northern Ireland's lack of international legitimacy affects the balance of forces in the conflict partly directly and partly by its impact on internal perceptions of the balance. On the one hand, it places powerful constraints on the government's response to the violence because its vulnerability to criticism for human rights violations is compounded by the problem of Northern Ireland's status, and, on the other hand, it gives credibility to the notion that the minority in Northern Ireland can defeat the majority. Allied to this is a belief that the international community would never permit the minority to lose and might just conceivably assist it to win.

As a result of the operation of these factors the Provisional IRA is able to present a threat to the existence of Northern Ireland that has some measure of credibility despite the fact that the organisation's potential support is limited to the minority and that it has actually been supported only by a minority of the minority. The tendency to overestimate the strength of the forces on the nationalist side has another effect. It underpins the assumption that a united Ireland is bound to follow a British withdrawal.[23] The significance of this is that it enables the Provisionals to argue that they can be successful without winning the support of Protestants.[24] In fact, given the relationship between violence and sectarian divisions in Northern Ireland, it would be extremely difficult to sustain the credibility of a campaign of violence that depended for its success on support across the community divide. The assumption that Protestant opposition by itself cannot prevent the achievement of a united Ireland therefore forms a crucial element in

Republican calculations.

Finally, Northern Ireland's lack of international legitimacy lies behind the persistent ambiguity in constitutional nationalism over the meaning of the commitment to seek unity by consent. On the one hand, there is a broad consensus among constitutional nationalists condemning the use of violence to achieve unity and some recognition that partition was the product of divisions and not just their cause. On the other hand, there is a reluctance to adopt any position that might be interpreted as according legitimacy to Northern Ireland as a political entity. The reluctance is understandable since the denial of legitimacy to Northern Ireland attracts international support for the nationalist case. At the same time, the ambiguous position of constitutional nationalists has fuelled Unionist suspicion of their intentions and has consequently been an obstacle to the achievement of political accommodation within Northern Ireland that constitutional nationalists desire. Unionist suspicion has not been allayed by the commitments entered into by the Republic of Ireland under the Anglo-Irish Agreement, because the assurances on the question of consent have been given in the context of a process that Unionists see as designed to achieve the result of a united Ireland. Attempts by British ministers to reassure Unionists by suggesting that the Republic's commitments in practice entail the permanent acceptance of the Border have simply made matters worse because they have prompted nationalist protests that this is not in fact the case.

The overall impact of the internationalisation of the conflict has been to reinforce the fundamental role that the threat to Northern Ireland's existence as a political entity plays in entrenching the province's divisions. Because it is difficult to disentangle the external from the internal causes of the sectarian divisions in Northern Ireland, it remains open to argument just how much weight should be placed on the international dimension as a factor in the conflict. The thread connecting internal and external causes of the conflict is the question of legitimacy, particularly as it relates to the legitimisation of violence.

The efforts of the British government during the troubles have for the most part been directed towards the achievement of political accommodation between the constitutional parties within the province. However, a significant weakness of the various schemes that have been put forward for an internal settlement has been their incapacity to resolve the question of Northern Ireland's place in the world. The Anglo-Irish Agreement, which gives the Republic of Ireland an input into British policy in Northern Ireland through an intergovernmental conference, is very far from providing an answer to this question, but its success in attracting international support has reduced the concern over Northern Ireland's anomalous status in the outside world and thereby helped to contain the conflict. The Agreement has also been a factor in checking the electoral rise of Sinn Fein, the political wing of the Provisional IRA, though the threat from the Provisional IRA has by no means diminished to the point where Unionists might see benefits in the Agreement. Indeed, it seems unlikely that the violence of the Provisional IRA will end as long as sectarian divisions continue to be reinforced by the uncertainty that exists over the province's future as a political entity.

Moreover, the view that the Agreement has significantly improved the prospects for an agreement among the principal local parties, excluding Sinn Fein, is hard to sustain. The greater flexibility being shown by Unionists in relation to proposals for power-sharing is much more a reflection of their desperation to find a means to negotiate away the Anglo-Irish Agreement than it is an indication of a lessening of sectarian tensions that would be capable of sustaining such a settlement. Furthermore, the emphasis placed on the issue of power-sharing is misleading in so far as that implies that lack of agreement on power-sharing is the crux of the difficulty in reaching a settlement. At least as important to the parties is the question of whether a settlement appears to advance or retard the prospects for a united Ireland. This brings us back to the fundamental problem of Northern Ireland's anomalous status and the difficulty of building anything on such shaky foundations. The great virtue of the Anglo-Irish Agreement has been its durability. It is important that the search for better solutions should not endanger what has already been achieved through the Agreement, limited though that is. Further, in fairness to the province's politicans, the difficulty of conducting negotiations in the circumstances of Northern Ireland's lack of international legitimacy needs to be recognised.

In South Africa's case it is not the territory or its boundaries that lack international legitimacy but the nature of its government. Indeed, the international community has insisted on the legitimacy and permanence of the country's 1910 boundaries. Consequently the homelands given independence under the policy of apartheid have failed to secure international recognition. Self-determination has been interpreted as the right of the majority to establish an independent state in any area administered as a political entity by a colonial power. However, the international community has not accepted the notion that the homelands constitute colonial territories in relation to South Africa. Hence, from the perspective of international opinion, the decolonisation of

Lesotho constituted a legitimate application of the principle of self-determination, while the creation of Transkei was an illegitimate denial of the right of the majority within South Africa, notwithstanding some resemblances between the two entities.

South Africa's conflict with the principle of self-determination derives from the exclusion of the majority of its population from its central political institutions and of course, unlike Northern Ireland, the issue has been formally taken up by the international community. It is also clear what the international community prescribes as a solution and that is one person one vote in a unitary state. In this context, a unitary state means essentially an undivided South Africa. In other words, it is arguable that federalism *per se* would not run counter to the prescription. However, federalism as a device for limiting the effective political rights of the majority almost certainly would be seen as a denial of self-determination. If this appears to leave considerable room for argument as to what does or does not limit effective rights, in practice the genesis of any constitution is likely to be more important in the international judgement of it than its precise provisions. In this connection, it is important to distinguish between two processes, the external application of the principle of self-determination and what Jack Spence has usefully described as the "right of internal self-determination".[25]

Under the norm of non-intervention as enunciated in a UN Declaration on the subject in 1966, "every state has an inalienable right to choose its political, economic, social, and cultural system, without interference in any form by another state".[26] The catch is that compliance with the norm of self-determination remains a pre-condition for the exercise of this choice. Thus a constituent assembly elected by universal suffrage in internationally accepted elections would be in a position to write elaborate safeguards for minorities into a constitution, if it so chose. Equally it could opt for a federal system of government. Namibia provides the obvious topical example of the creation of a constituent assembly that allows for such options.

Perhaps a way forward for South Africa itself might eventually also lie in the election of a constituent assembly on the basis of universal suffrage, but requiring an even larger weighted majority (say, 80 per cent of the representatives rather than two thirds as in Namibia) to make decisions. Such a constituent assembly could provide the setting for a process of political bargaining, the successful outcome of which would be assured a large measure of both external and internal legitimacy. In any event, it is clear that safeguards for minorities could not be written into a constitution in advance of a constituent assembly and the ensuing bargaining without jeopardising the legitimacy of the transition. This would be the case even if the government had arrived at the safeguards through a process of negotiations with elements it regarded as representative of the majority of the population.

This should not be regarded as simply a piece of theoretical hair-splitting. It is of real practical importance, as the case of Zimbabwe-Rhodesia showed. The internal settlement between Ian Smith and Bishop Abel Muzorewa went much further in the direction of unadorned majority rule than anything currently being contemplated by the South African government. However, despite the participation of a majority of the electorate in the internal elections in 1979 and a favourable report by British Conservative Party observers on the conduct of the poll, the incoming Conservative government in Britain declined to recognise Zimbabwe-Rhodesia. Its refusal was on the grounds that Zimbabwe-Rhodesia's constitution contained provisions entrenching white privileges that the majority had not given their consent to. The consequent failure of the internal settlement to achieve international recognition led to British mediation that came to fruition in the Lancaster House Agreement.

While the Agreement eliminated some of the protection for whites in the Zimbabwe-Rhodesian constitution, it also stopped short of imposing unadorned majority rule on the country. However, the Lancaster House Agreement had the advantage of being seen as involving all significant elements of black opinion in Zimbabwe and of enjoying the backing of the Commonwealth. Another significant difference between the internal settlement and the Lancaster House Agreement was simply that the former failed to stop the war. But it would be wrong to see this as divorced from the constitutional issue. The lack of international legitimacy of the regime created by the internal settlement was a factor in the continuation of the war. More generally, all systems characterised by coercive dominant-subordinate relationships that are no longer able to secure the dominant community's position by force face the problem of how to criminalise (in other words, marginalise) political violence from the subordinate community. Concessions to the subordinate community that fail to meet international expectations of what constitutes a legitimate settlement are unlikely to end political violence. Hence external perceptions that the conflict remains unresolved are likely to persist; perceptions that themselves will influence the internal parties.

Perhaps the most important lesson for South Africa from the case of Zimbabwe-Rhodesia is the desirability of external mediation. Deviation from a literal interpretation of one person one vote in a unitary state is most likely to secure international acceptance when it is clearly seen to be the outcome of a process in which the international community has been involved or which it has endorsed. This applies even in the case of the election of a constituent assembly.

Namibia's route to independence provides a striking example of successful mediation by the international community. The key role was played by five Western members of the Security Council. Their opposition undermined South African efforts to sustain shaky internal settlements. At the same time, their insistence in the 1982 negotiations that the writing of the constitution should require a two-thirds majority played an important role in gaining eventual white acquiescence in the transition to majority rule. Admittedly, in both the case of Namibia and Zimbabwe, external mediation was buttressed by the legal status of the territories. Such a justification for external mediation is absent in South Africa's case. What states might act as mediators is therefore unclear.[27] What is clear is that such mediation would, in practice, require the co-operation of the leading Western powers, the Frontline states, and probably the Soviet Union as well.

Despite the advantages of external mediation, it is not difficult to see why dominant communities are attracted to the notion of an internal settlement. Firstly, an internal settlement appears to provide a setting in which the influence of international prescription can be minimised. Secondly, it provides a context for bargaining in which what matters most is the local balance of forces. (The Unionists in Northern Ireland have justified their refusal to negotiate with the SDLP while the Anglo-Irish Agreement remains in force on the grounds that the very existence of the Agreement puts them on "unequal" terms with the minority.) Since the local balance of forces must favour the dominant community for its position as a dominant community to be viable at all, this places the issue of broadening political participation, sharing power, or making some other concession to the subordinate community, in a framework that gives most reassurance to members of the dominant community.

Finally, an internal settlement tends to be seen by the dominant community as a context in which external influences on the subordinate community are minimised and hence one in which representatives of the subordinate community are able to display greater realism. The record of the subordinate community's past acquiescence in the rule of the dominant community is seen as providing ample evidence that support will be forthcoming from the subordinate community for such realism. The dominant community's hope is that even if an internal settlement fails to bring to an end the system's lack of international legitimacy, its effectiveness as a basis for the maintenance of internal political stability will secure international acquiescence, much as the dominant community's exercise of a monopoly of power once did.

In practice, however, internal settlements face formidable obstacles to their success. Usually, dominant communities only make concessions to a subordinate community whose political aspirations are supported externally in circumstances where they lack the capacity to continue to maintain their effective monopoly of political power. In short, such proposals tend to emerge in a crisis. This brings us back to the issue of violence.

While the lack of international legitimacy of white minority rule in South Africa is not in doubt, the question of the legitimacy of the use of violence against that rule is more complicated, especially in relation to Western opinion. This is partly because of the extent to which Western societies identify whites in South Africa as a fragment of the West itself. Robert Jackson uses the notion of cultural proximity to describe the relationship. On the one hand, this fuels demands within Western societies that the Republic be made to conform to current Western standards in relation to racial discrimination. On the other hand, as Jackson points out, "cultural proximity rules out military intervention by the West in South Africa".[28] It also inhibits support for groups engaged in organised "armed struggle". Thus unusually in the case of South Africa, the issues of the legitimacy of the regime and its policies and the legitimacy of anti-system violence appear to have been disaggregated.

Normally, attitudes towards revolutionary violence against a regime tend to be a function of whether that regime is seen as legitimate or not and there is generally a high measure of tolerance of, if not support for, violence against a regime that is perceived to be illegitimate for whatever reason. An obvious example is provided by Western attitudes towards the regime in Kabul. Since the armed intervention of the Soviet Union installed a new regime in Afghanistan in 1979, the West has supplied massive quantities of weapons to its opponents without the slightest public controversy being generated over the legitimacy of the West's counter-intervention. Only now with the withdrawal of Soviet troops is the question of whether it is in the West's interests to continue to arm Islamic fundamentalists beginning to be aired.

In the South African case, however, there are definite Western inhibitions on the use of violence against white minority rule. This forms an important, if generally unstated, parameter in the debate on sanctions. The effectiveness of Archbishop Tutu's advocacy of sanctions almost certainly owes much to the emphasis he places on sanctions as an *alternative* to violence. An escalation of organised violence against the regime would limit the appeal of this argument. Therefore, quite apart from the strategic difficulties in the way of "armed struggle", it is arguable that the ANC has an interest in limiting the scope of such attacks as long as it considers that there is more to be gained from a tightening of economic sanctions. It also follows that disillusionment in the subordinate communities in South Africa with sanctions as an instrument for bringing about political change may translate into increased support for anti-regime violence.

The test by which sanctions will be judged (and it does not matter for this purpose what their actual impact has been) is whether negotiations towards a settlement satisfying existing international norms prove possible. This is an extremely tall order given the incongruence that exists between what international norms prescribe and the actual balance of political forces within South Africa itself. Admittedly, South Africa's problem is not quite as severe as that of Northern Ireland in this respect. It is hard to envisage the balance of forces ever changing sufficiently in Northern Ireland to make possible the united Ireland most of the world appears to regard as inevitable. In South Africa's case, majority rule does at least seem inevitable in the long run. However, the exaggerated expectations of its imminence in the outside world seem likely to exacerbate the difficulties of the transition period by burdening black representatives with unachievable timetables for change, while at the same time maximising the likelihood of white backlash in response to political concessions by the government.

The case of Israel has similarities with both that of South Africa and that of Northern Ireland. However, there is also a very significant difference. The international community played an important role in the creation of the state of Israel through the UN partition plan. Consequently, notwithstanding the hostility Israel has faced in the Middle East, as far as the wider international community is concerned there is no question that Israel constitutes a legitimate political entity, endowed with all the rights of a sovereign state, including the right of self-defence. What remains at issue is the question of the legitimate boundaries of the state.

The general consensus of world opinion expressed through the UN seems to be that the boundaries of Israel prior to the war of June 1967 constitute the rightful limits of the state. However, the UN Security Council resolution generally seen as embodying the principles of a settlement of the problem, that is to say resolution 242 of 1967, is by no means unambiguous on this point. It called for Israel's "withdrawal from territories occupied in the recent conflict", while proclaiming the right of every state in the area "to live in peace within secure and recognised boundaries free from threats or acts of force".[29] That leaves scope for the Israeli government to argue that its right to secure boundaries rules out a complete withdrawal from the territories occupied during the 1967 war and that the absence of the definite article before "territories" in the resolution provides support for this interpretation.

However, if one leaves the ambiguity in resolution 242 to one side, it is clear that there has been a hardening of international opinion, which has been reinforced by the *intifada*, that Israeli rule over the West Bank and Gaza strip is illegitimate. The fact that Israel has failed to annex the territories (apart from East Jerusalem) inevitably makes an issue of their status in a world that has embraced the sovereign independent state as the model of the legitimate polity leaving scant room in practice for alternative modes of implementing the principle of self-determination. Therefore it is not surprising that there is also a wide international consensus that Israeli rule should be replaced by the establishment of a Palestinian state in the occupied territories.

Admittedly, the readiness of the United States to push for a two-state solution remains in doubt. Much rests on the United States government, since it is perceived rightly or wrongly as having greater influence over Israel than the rest of the world put together. Arafat's acceptance last December of resolution 242 and of Israel's right to exist and his renunciation of all forms of terrorism, thereby meeting American conditions for a dialogue between the United States government and the Palestine Liberation Organisation, is a gamble that the combination of American pressure and the *intifada* will succeed in shifting the Israelis. It is of course critically important to this strategy that the *intifada* itself not be seen either as embodying a form of terrorism or even as permitting terrorism as one of a number of methods of resisting occupation.

This part of the gamble seems likely to be successful. There are, in fact, a number of reasons why Israeli attempts to portray the *intifada* as a continuation of terrorism are likely to fail. Firstly, the West is primarily concerned with Palestinian involvement in *international* terrorism, not with political violence within the greater Israel. Secondly, the coldly calculated atrocities that Western public opinion associates with terrorism are far removed from the stone-throwing and other spontaneous acts of violence that appear to embody the *intifada*.

Thirdly and perhaps most importantly of all, the actions of the Israeli army in attempting to quell the *intifada* have cast the Palestinians in the role of the primary victims of violence. Unless there were to be a dramatic change in the numbers killed on each side in the conflict, that seems unlikely to change.

On this last point, there is some similarity between the *intifada* and the height of the township unrest in South Africa between 1984 and 1986. The enormous international damage the South African government suffered during this period owed much to the fact that the dominant image of these events was of state violence against blacks. (The obvious topical comparison is with the suppression of the pro-democracy movement in China.)

The same argument may be applied to Northern Ireland. The British government suffered very considerable damage to its reputation during the hunger strike by Republican prisoners during 1981, precisely because they appeared to outside opinion to be the victims of a dubious policy.[30] Indeed, it is quite evident that much greater sympathy has been generated both domestically and internationally for the Irish nationalist cause by martyrdom than by insurgency or terrorism. That is not to say that movements such as the Provisionals have deliberately pursued martyrdom as a policy. That option seems by and large to be confined to groups such as Islamic fundamentalists who possess a religious sanction for martyrdom.

However, the capacity of the *intifada* to generate international support for the Palestinians does not simply derive from the picture it presents of Palestinians as victims of violence. The fact that it appears completely divorced from any threat to the existence of Israel proper from neighbouring states is also crucially important. Thus, at one and the same time, the *intifada* has demonstrated the strength of opposition to the continuing occupation of the West Bank and Gaza and cast doubt on the relevance of the occupation to Israel's own defence, the only justification there is in terms of international norms for Israel's presence in the territories. Indeed, the more Israel has appeared to be an example of a divided society beset by domestic conflict, the less sympathy there has been in the West for the Israeli cause.

What remains to be seen is whether increasing international isolation provides a more potent influence on Israeli policy towards the Palestinians than the external military pressure of Arab states did in the past. In particular, has the diminution of the military threat to Israel paradoxically enhanced the prospects for a Palestinian state? In this context, one can look at South Africa's total strategy as an attempt, which largely failed, to portray the Republic in the same light as Israel tended to be seen prior to the peace treaty with Egypt, that is to say, under siege from anti-Western forces.

In the case of Northern Ireland, the British government has tended to play down the external aspect of the conflict, preferring to present the conflict as a sectarian one between Protestants and Catholics rather than as a battle between Unionists and nationalists over the province's political destiny. In particular, Unionists have failed in their efforts to persuade British governments that the irredentism of the Irish Republic constitutes the root of the conflict. The domestic perspective is less embarrassing to Britain because it focuses attention on the prejudices of the two communities in the province, placing the responsibility for the continuation of the conflict squarely on the local population, whereas the external perspective highlights Northern Ireland's anomalous status within the United Kingdom, widening the field of responsibility as well as underlining the intractability of the situation.

Where Israel appears to be fundamentally different from both South Africa and Northern Ireland is that in Israel's case the dominant community has the option, through the two-state solution, of a settlement that promises not merely to keep the community's control over the political institutions it has created, but to sustain its lifestyle intact, with the possible exception perhaps of Israeli settlers on the West Bank. Furthermore, there would appear to be every prospect that such a settlement would prove durable, given the international support it could expect.

By contrast, the option probably most readily available to the dominant community in Northern Ireland, power-sharing in the context of a devolved government, suffers from having been tried once and failed. Its failure was largely due to the unpopularity among Protestants of the Sunningdale Agreement of December 1973, which linked the power-sharing experiment with a Council of Ireland to promote co-operation between the two parts of Ireland. While it is unlikely that such a settlement would have a major impact on the lifestyle of the dominant community, the prospect of such a settlement does not enthuse the dominant community, not least perhaps because of that community's preoccupation with Northern Ireland's constitutional status, an issue such a settlement would not address directly or fix permanently. In particular, such a settlement could not end the aspirations of nationalists, North and South, to achieve a united Ireland. Furthermore, given the likelihood also of the continuation of Republican violence in these circumstances, it may be misleading to describe such an arrangement as a "settlement" at all.

In South Africa's case, much greater sacrifices in terms both of the loss of political power and of changes in lifestyle appear to be required of the dominant community to break the current political impasse. That is probably a good reason in itself for predicting the persistence of the impasse. Yet superficially at least the appearance of political movement and desire for political accommodation seems stronger in South African than in either Israel or Northern Ireland. A possible explanation is that South Africa remains less affected than the other two societies by inter-community violence, paradoxically because the much greater level of social and economic inequality in South Africa has reduced the scope for such conflict.

Notwithstanding the differences among them as societies, as political systems Northern Ireland, South Africa, and Israel all lack international legitimacy because of conflict with the current interpretation of the principle of self-determination. Given the stake that most states of the present international political system have in the current interpretation of the norm, change in its interpretation to accommodate the status quo in the three societies remains extremely unlikely. Furthermore, in each case hostility towards the norm is confined to the dominant community. For the subordinate communities, the support of the international community represents for the most part a source of hope. The dominant and subordinate communities consequently tend to be divided in their attitudes towards external intervention stemming from the system's lack of legitimacy. Because such external intervention rarely has the capacity to effect a significant shift in the local balance of forces, arguably it is generally more likely to contribute to political polarisation than to provide the impetus for political evolution.

However, at the stage of political impasse, external mediation may provide a way forward for these societies because it offers a context in which international support can help to legitimise settlements that are the product of negotiation rather than simply international prescription. At the same time, it is important to recognise that external mediators generally have a very limited capacity to force settlements on the internal parties. In particular, without the clear consent of both the dominant and the subordinate communities, no question of external military intervention to back up a settlement could possibly arise. International peacekeeping is a matter of policing agreements, not a substitute for their absence. External mediators may be in a position to put economic pressure on the parties to settle, though, in practice, economic leverage tends to be a somewhat blunt instrument with which to effect political changes.

ENDNOTES

1. Robert Jackson: "Negative Sovereignty in Sub-Saharan Africa", *Review of International Studies* Vol 12, No 4, October 1986, p255.
2. Wight defined international legitimacy as "the collective judgement of international society about rightful membership of the family of nations; how sovereignty may be transferred; and how state succession is to be regulated, when large states break up into smaller, or several states combine into one". Martin Wight: *Systems of States* (edited by Hedley Bull), Leicester, Leicester University Press, 1977, p153.
3. A factor that compounds Israel's lack of legitimacy in respect of the territories is that from the perspective of international law, Israel's policy of settlements in the West Bank and Gaza constitutes a violation of the 1949 Geneva Convention on military occupations. The issue is clearly explained in Adam Roberts: "Decline of Illusions: The Status of the Israeli-occupied Territories over 21 Years", *International Affairs*, Summer 1988.
4. This superseded the 1960 Declaration on the Granting of Independence to Colonial Countries and Peoples (General Assembly Resolution 1514 (XV), 14 December 1960).
5. General Assembly Resolution 2625 (XXV), 24 October 1970. 25 UNGAOR, Supp. 26 (A/8026) p124.
6. M Wight, op. cit., p168.
7. UN General Assembly Resolution 3314 (XXIX), 14 December 1974.
8. There are some examples of secession by mutual agreement between a central and a regional government or between two governments forming a loose confederation. The most important case was Singapore's secession from the Malaysian Federation in August 1965.
9. Michla Pomerance: *Self-Determination in Law and Practice*, Martinus Nijhoff, The Hague, 1982, p1. The issue of whether self-determination constitutes a *jus cogens* is discussed in Pomerance, pp63-72.
10. For example, according to Leo Kuper, South Africa, Israel, and Chile "predominate overwhelmingly" in the proceedings of the United Nationas Commission on Human Rights and the Sub-Commission on the Prevention of Discrimination and Protection of Minorities. Leo Kuper: *Genocide*, Harmondsworth, Penguin, 1981, pp161-5. Amnesty International is universal in its approach, but its reports on certain countries have a greater political impact than others because of judgements on their legitimacy. For its contribution on Northern Ireland, see *The Protection of Human Rights in Northern Ireland*, London, Amnesty International, 1985.
11. See, for example, the British Prime Minister's interview with *Newsweek*, 16 May 1983. "Northern Ireland is free to determine its own future. It is a fundamental part of the United Kingdom. If the majority of the people in Northern Ireland wish not to be, obviously we would honour their wish, whether it was to be independent or to join up elsewhere. Northern Ireland is part of the United Kingdom because of the wish of the majority of its people."
12. Report drawn up on behalf of the Political Affairs Committee on the situation in Northern Ireland. Rapporteur: NJ Haagerup, 19 March 1984 (European Parliament Working Documents 1983-84, Document 1-1526/83), p37.
13. See the chapter on "European Interventions" in my book, *Northern Ireland: The International Perspective*, Dublin, Gill and Macmillan, 1988.
14. See, for example, Phillip Knightley: "Is Britain Losing the Propaganda War?", *The Sunday Times*, 31 May 1981.
15. Lloyd George appreciated the weakness of the British case on partition in 1921. He outlined his strategy on the negotiations with the Sinn Fein leaders to his cabinet colleagues in the following terms: "If the conference started without securing in advance Irish allegiance to the Crown and membership of the Empire, the discussion would become entangled in the Ulster problem; that De Valera would raise the

question of Fermanagh and Tyrone, where we had a very weak case, the conference might break on that point, a very bad one. He would rather break — if there was to be a break — now, on allegiance and Empire." Thomas Jones: Whitehall Diary: Ireland 1918-1925 (ed) Keith Middlemass, London, Oxford University Press, 1971, p111.

16. Hispaniola, Tierra del Fuego, Borneo, New Guinea, St Martin, and Usedom (on the border between East Germany and Poland) are the only current cases of islands under divided sovereignty besides Ireland. To this list may be added the *de facto* partition of Cyprus and the presence of an American naval base on Cuba.

17. See "Arms and the Man: A dealer talks of South Africa's missile links with Belfast", *Sunday Telegraph*, 9 July 1989.

18. Ibid.

19. See the chapter on "The American Connection" in my book cited in note 13 above.

20. See the results of the survey in Edward Moxon-Browne: *Nation, Class and Creed in Northern Ireland*, Aldershot, Gower, 1983, p40.

21. See the survey results in *The Irish Times*, 12 February 1986.

22. The remarks made by Ireland's Ambassador to Britain, Dr Eamon Kennedy, to the British Prime Minister, Mrs Margaret Thatcher, in December 1979 provide a striking example of nationalist hopes. He said: "Prime Minister, it's Christmas and you've given us the gift of peace in Zimbabwe. It must have demanded great energy and vision and courage to achieve that. Now could we ask that having achieved that in Africa, you might channel those same qualities towards an island next door, where our tribes need to be brought together too?" Interview with *The Irish Times*, 15 July 1983. On Loyalist fears, see as an example the comments in the UDA's journal, Ulster (Belfast), February 1984.

23. Conor Cruise O'Brien's dire warnings over the years that a British withdrawal would be followed by civil war, repartition, and the emergence of an independent Loyalist state within truncated boundaries have made little impression on nationalist opinion on this issue.

24. See, for example, the remarks of a Sinn Fein local councillor in "SF, communists debate socialism", *The Irish Times*, 21 September 1987. He argued that "until the British military presence in Ireland was smashed the Protestants could not be won over to the national struggle". The line taken by the communist speaker was that they had to be won over and not surprisingly he criticised the Provisional IRA campaign.

25. J E Spence: "The Most Popular Corpse in History", *Optima*, Vol 34, No 1, March 1986, p16.

26. UN General Assembly Resolution 2131 (XX) of 1966.

27. See R W Johnson's discussion of what states might constitute a "collective metropole" for South Africa in "How Long *Will* South Africa Survive?", *Die Suid-Afrikaan* (Cape Town), February 1989.

28. R Jackson: "Negative Sovereignty", op. cit., p260.

29. Quoted from United Nations Security Council Resolution 242 "Concerning Principles for a Just and Lasting Peace in the Middle East", 22 November 1967.

30. The policy of criminalisation was an effort to isolate those engaged in political violence in Northern Ireland. It was linked to the ending of detention without trial. It was initially quite successful. However, the more prisoners were able to generate public support for their demands that their political motivation should be recognised in the way they were treated as prisoners, the more dubious the policy naturally appeared.

6 The peace process and intercommunal strife

MERON BENVENISTI

It is a safe assignment to serve as an analyst dealing with the Middle East. One's gloomy forecasts are almost invariably confirmed and even perceived as quite cheerful: everyone expects even worse news.

It is also a permanent job, as nothing seems to change and no solution is in sight. Issues raised twenty, forty years ago seem valid forever. So one does not have to alter one's messages as they seem perennially relevant; nor should one suggest new remedies. The problem, once defined, remains unaltered and the paradigm durable. After all, we are no closer to resolving the conundrum, so why bother to redefine it?

One reason for this conservative approach to Middle Eastern problems, besides intellectual laziness, is the blinding blaze of the Israeli-Arab conflict. This perpetual conflagration dominates the horizon and eclipses all other quarrels which tear the region apart. The world is transfixed by the old tragedy of Jew and Arab fighting for possession of the Holy Land. So profound is this tragedy that it dominates the Middle Eastern scene and subsumes all its malaise.

Nobody can remain indifferent to the Israeli-Arab conflict. The drama of Israel rising from the ashes, the two thousand years of Jewish-Gentile encounter, the plight of the Palestinians who have to pay for the wickedness of others, the image of the quintessential victim turned oppressor, the Biblical association — all evoke powerful emotions which make this strife a clash of ideological and spiritual shibboleths rather than an international conflict.

So powerful is the hypnotising effect of the Israeli-Arab conflict, that one pays little attention to fundamental changes that have transformed the situation in the Middle East since it first appeared on the international agenda. The almost axiomatic conviction that the Israeli-Arab conflict is the core trouble of the region remains an ideological percept but has been long overtaken by events. The perception that the fertile crescent is still in the era of wars that change geo-political facts is the premise prompting diplomats and politicians to engage in a peace process. Yet this premise is itself obsolete after Camp David, the Israeli-Egyptian peace treaty, the Lebanon campaign of 1982, the *intifada* and Jordan's disassociation from the West Bank.

These events ushered in a new phase for Israel and its neighbours that can be called unstable equilibrium. It is an unstable yet durable geo-political system maintained by Israel and Syria. Each needs the other as an enemy. Each, for its own reasons, needs the conflict to continue unresolved but in a manner that will not explode into all-out war. This equilibrium is facilitated by the cold yet durable peace between Israel and Egypt, the vulnerability of the Hashemite regime in Jordan and the profound impact of the aftermath of the Iran-Iraq war.

The notion that the Palestinian cause can serve as a battle-cry that will shatter the unstable equilibrium is maintained only by a minority of ideologically motivated observers in Israel and in the West, who in their despair believe in a *deus ex machina* that would resolve the old Jewish-Arab feud. The Palestinians themselves have given up that hope. The Arab states care very little about the Palestinians, as was demonstrated quite clearly during the Lebanon campaign and during the long months of the *intifada*. Arab regimes believe they have done what they could for the Palestinians and maintain involvement in the Palestinian cause only when it serves their interests.

The Israeli-Arab conflict, which for forty years has been a region-wide, inter-state conflict, has shrunk to its original core, namely Israeli-Palestinian intercommunal strife. This major turning point, which renders the traditional paradigm obsolete and requires reformulation of options and choices, is not acknowledged by most observers. It is precisely in perceiving the reality of Israeli-Palestinian relationship at the end of the 1980s that denial, evasion and ideological catechism are most powerful.

To understand the new phase in Israeli-Palestinian relationship one must return to the formative phase of this tragic encounter. To be sure, one can fix the starting point at the end of World War I or even earlier, in 1882 when the first Zionist settlement was established. But it seems that the mid-1930s, and more specifically the Arab Revolt of 1936 and its aftermath, have defined the contours of the dispute and formed the point to which we can trace back present relationships, perceptions and even strategies.

By the mid-1930s both the Jewish and Palestinian communities had developed into cohesive and self-sustaining societies and were moving irrevocably and consciously towards total confrontation. The Palestinians, aware of their growing national power, endorsed "armed struggle" as their strategy and launched the Arab Revolt, a tremendous effort to overthrow the British Mandate and destroy the Zionist enterprise. Their primary targets were the British because they viewed the Zionists as white-settler colonists, totally dependent on Britain and bound to disappear once the colonial power was ousted.

The Zionist reaction was as powerful. Even the most moderate amongst them understood that a bloody showdown was inevitable. They had to abandon their naive and self-serving perception of the conflict as an international class struggle, or as a tragic misunderstanding caused by the ignorance of the natives, who would learn to accept the Jews because of the material benefits they brought. The Zionists realised that the Palestinian cause was in fact a national movement but, unable to grant it legitimacy, depicted it as a fascist, reactionary alliance of murderers. They viewed the Palestinians as a nonviable society, an offshoot of the Arab world, not an independent factor.

After 1936 the perceptions of both Palestinians and Jews were characterised by exclusive attitudes, with the conflict perceived as a zero-sum game and an externally generated dispute. Each side ignored or underestimated the other, viewing it as an object manipulated by external forces. As a result, they saw no point in trying to relate directly, opting rather to deal with the external forces perceived to be in control.

The defeat of the Palestinians in 1937/38 prompted neighbouring Arab countries to adopt their cause, which confirmed Jewish perceptions of the Palestinians as an externally generated force. The 1948 war reinforced these perceptions. The collapse and physical destruction of Palestinian society completed the process of externalisation, as objective reality caught up with perceived reality. The intercommunal strife became a conflict between sovereign states — the Israeli-Arab conflict.

The Palestinians themselves assisted the Israelis in redefining the conflict. During the Pan-Arab, Nasserite era Palestinian activists perceived their national struggle in the broader, anti-imperialist context. They clung to their old perceptions, viewing the "Zionist entity" as a neo-colonial, nonviable phenomenon, relying for its sheer survival on imperialist power.

The 1967 war, the occupation of the West Bank and Gaza, the 1973 war, did not change these perceptions, which had become fundamental credo. They persisted despite the gradual elimination of external, interstate disputes, culminating in the signature of the Israeli-Egyptian peace treaty. They have not changed despite the exacerbation of the intercommunal strife between Jews and Arabs in the occupied territories, for they satisfy deep-seated psychological needs and enable both sides to believe in the exclusivity of their claims.

The inability of each side to accept the legitimacy of the other even as an enemy, let alone as a partner for peace negotiations, is central to the understanding of the failure of traditional diplomacy in its attempt to resolve the Israeli-Palestinian dispute. The diplomatic vocabulary is not designed to cope with fundamental issues of self-identity, self-expression, existential fears of annihilation, clash of symbolic interests, absolute justice. Intercommunal disputes of such proportions are beyond diplomacy, and therefore attempts to resolve them through the traditional "peace process" approach are bound to fail.

In general terms, the peace process is a linear, means-end effort to transform war situations into peaceful conditions. It is designed to formulate answers to the questions which result from a clash of national interests in the international arena. In order for the peace process to be effective, certain conditions must be met.

First, it must take place within the context of an international system. This system is based on recognition of its national members as legitimate, independent actors who may interact with other members on an equal basis. Participants in the peace process must be perceived as accredited entities, who represent extraneous power structures over whom other members have no authority, and who report to independent constituencies. The issue for discussion is not the basic right of the enemy to an autonomous and separate identity; it is the circumstances under which that right is exercised. It is only within this procedural context that the peace process can function.

Substantive issues have their own constraints. The peace process must be premised on the assumption that the conflict is not a zero-sum game. The belligerents must be prepared to participate in negotiations in which concessions and compromises can take place. Issues must be translatable into clearly defined texts and must focus on concrete areas of dispute.

It is only because all those conditions were met that the peace process between Israel and Egypt was successfully concluded. It is because none of these conditions can be met that the Israeli-Palestinian dispute cannot be resolved by a similar process.

In Security Resolution 338, both Israel and Egypt agreed on a means-end effort to transform hostilities into peaceful relations, within the context of the international system. Sadat's visit to Jerusalem symbolised the recognition of Israel as a legitimate actor and the Israel public as an autonomous and independent constituency. Secret Israeli-Egyptian meetings prior to official negotiations established a positive-sum game: the return of Sinai and an unconditional recognition of Egyptian sovereignty in exchange for security arrangements. Bilateral issues could be translated into a clearly defined text, and disputed issues could be broken down into concrete items on which concessions and compromises could be reached.

The Israeli-Palestinian dispute is stuck at the critical pre-procedural phase. Although objective observers would like to define the dispute as one involving a clash of national entities struggling for the same land, this definition is not accepted by the adversaries. The core of the conflict is understood by them as "survival". The struggle goes beyond the apparent physical survival of the peoples involved and encompasses basic issues of identity and integrity. The core of this issue is therefore non-negotiable, for issues of identity are a zero-sum game.

Furthermore, we pointed out that the peace process must take place within the framework of the international system. In order to do so, however, both members must be full members of that system. Negotiations take place around tables with representatives of the belligerent parties equipped with their symbols of legitimacy. Diplomats present credentials, flags are displayed and national anthems respected. Yet the very core of the Israeli-Palestinian conflict is over the legitimacy of these symbols, the very right of each other to exist. Participation itself therefore implies a symbolic concession too great to make.

Moreover, the status of Israelis and Palestinians is asymmetrical. Israel is recognised as a sovereign state and her legitimacy is not disputed by most countries (including some Arab countries). The Palestinians are internationally recognised as a quasi-national entity but that status is too ambiguous to allow them participation in the international system as equal partners.

Despite its inapplicability, however, the diplomatic approach must continue because "international disputes" are handled by negotiators.

The only common language diplomats share is their particular jargon. Their only secure environment is "procedure", for diplomatic jargon is clinical and procedure is antiseptic. In that sterilised environment one converses in a dead language, the "Latin" of the Middle East: "242 and 338", international "forum", "umbrella", "conference", the Palestinian "issue", "freeze", "confederation", "working groups", "redeployment", "modalities for elections". Trusted envoys come and go, brilliant draftsmen produce working papers, secret conclaves convene, information leaks. It is the sub-culture of the mighty shrouded in the mystique of power.

"Something is going on," whisper the uninitiated, "it must be important, for otherwise important people would not be involved." But the real world speaks in the vernacular, and the environment of substance is polluted. The operation, alas, must be performed in field conditions; and in the field, alas, a bloody civil war is raging. Its arena is the entire area of former Mandate Palestine. Policies of enforced integration have completely erased the Green Line separating Israel proper from the occupied territories and created a new socio-political reality.

On the seventh day of the Six Day War the Second Israeli Republic was established in the area of former Mandate Palestine. It replaced the First Republic, which came into existence in 1948 and lasted nineteen years — a Jewish nation-state (with a small Arab minority), engaged in Jewish nation-building processes, and in armed conflicts with Arab states.

The political, societal, economic and administrative systems of the Second Republic took form gradually and consolidated twenty years later. Its government rules over all Mandate Palestine and has the monopoly on governmental coercive power in the entire area under its dominion. The distinction between Israel's sovereign territory and the area in which it rules by military government has lost its meaning. The Israeli government acts as sovereign, for all intents and purposes, in the whole area west of the Jordan river, changing the laws as it wishes and creating permanent facts.

The Second Israeli Republic on both sides of the obliterated Green Line is a bi-national entity with a rigid, hierarchical social structure based on ethnicity. Three-and-a-half million Jewish Israelis hold total monopoly over governmental resources, control the economy, form the upper social stratum and determine the educational and national values and objectives of the republic. The two million Palestinians divide into Israeli Palestinians and the Palestinians of the territories. Though the former — some 500 000 — are citizens of the republic, their citizenship does not assure them equality in law as one crucial test of citizenship is military, and Israeli Palestinians, who are exempt from service, are as a result second-class citizens.

The remaining one-and-a-half million Palestinians are citizens of a foreign state (in the West Bank) or altogether stateless (in Gaza). They are deprived of all political rights, ostensibly because they are under military occupation, but even their rights under international conventions governing military occupation are not assured, since the government of the republic does not recognise the application of these conventions to the territories.

The ethnic groups maintain economic interaction defined in professional literature as "internal colonialism", that is, the inferior ethnic group serves both as cheap labour and a market for finished consumer goods.

The Second Republic is by any objective standard a dual society and a political system whose technical term, again in professional literature, is "master-race democracy". The only reason this has not been universally acknowledged is that the territories have not been formally annexed.

Communal strife rages in the Second Israeli Republic. There is a perpetual conflict, not necessarily violent, between the Jewish majority group that seeks to maintain its superiority, and the Arab minority group (Israeli Arabs and Palestinians in the territories) that seeks to free itself from majority tyranny. The majority community perceives the struggle as one of "law and order". The minority community, which does not regard the regime as legitimate, seeks to destroy it. Both communities deny each other's standing as a legitimate collective entity. Hence, the Arabs define Zionism (the expression of the collective aspirations of the Jewish people) as racism — ergo illegitimate. The Israelis, in their turn, define Palestinian nationalism as Palestine Liberation Organisation (PLO) terrorism — ergo illegitimate. The delegitimisation is vital for both sides, for it enables them to believe in the exclusivity of their claim and in the absolute justice of their position.

Both communities, though internally divided, outwardly present a monolithic facade. The vast majority of the Jewish group are united in their aspiration to preserve the Jewish character of the Second Republic, that is, its superior status, even at the cost of democratic values. The distinction between personal-civic and ethnic-communal equality is accepted as conforming to liberal principles. The official (and legal) designation of the Second Republic as the State of the Jewish People, which implies second-class citizenship status for non-Jews, is the basic political creed of all Jewish (euphemistically defined as Zionist) parties. The differences between the main Jewish political factions are those of emphasis, style and abstract moral scruples but even these vanish utterly on the daily, tactical level.

The vast majority of the Arab group (Israeli Arabs and Palestinians) are united in their desire to destroy Jewish hegemony and are divided only over the most effective method for achieving that objective. Israeli Arabs wish to assert their place as a national minority and attain full communal equality with Jewish citizens. West Bank residents wish to secede and establish an independent state. Gaza refugees — stateless, landless — wish to return to their ruined villages and plundered lands and challenge the very existence of the Jewish republic. Just as all Israelis are Zionists, so too, all Arabs are supporters of the PLO.

The dynamic of the Israeli-Palestinian communal strife is similar to that of intercommunal strife everywhere — from Beirut to Belfast. It is waged in an endless cycle of violence, enforcement, domination, containment — fights over every piece of land, every tree. It is accompanied by the development of stereotypes, a lowering of the threshold of moral sensitivity, the loss of humanistic values, and despair leading ultimately to psychological withdrawal, to anarchy and fundamentalism. The conflict erupts periodically, usually after an unplanned provocation. Violence simmers just below the surface. It is an endemic condition, lacking a durable, ultimate solution.

The almost primordial nature of the Jewish-Arab strife has been masked by its international manifestations. Conceptual frameworks filtered facts and data, cozy answers sought facile questions and found them. It was too ominous to contend with an emerging intractable condition.

Then came the Palestinian uprising of December 1987, the symbolic declaration of the Israeli-Palestinian civil war. It should not have come as a surprise, least of all to the Israeli military. Israeli generals had monitored the gathering storm; all data relating to the intensification of spontaneous expressions of violent protest were at their disposal. They should have reached the same conclusions as independent researchers who warned that a new phase of Palestinian resistance and of the intercommunal strife had begun, a phase characterised by uncoordinated grassroots initiatives and carried out by angry young men acting on their own accord and undeterred by risks to their lives.

The dramatic increase of violent demonstrations since 1982 combined with a decrease in PLO-initiated terrorist acts should have sounded the alarm. But Israeli generals were captive to the obsolete conception of the externally generated conflict. The internal civil war caught them by surprise. December 1987 was Israel's second Yom Kippur. In 1973 they misread the signs of imminent external attack; in 1987 they ignored signs of imminent internal uprising. In both cases, panic seized the generals and blunders ensued, except that in 1987 it was the Palestinians who paid with their lives.

Israeli generals learn fast, however, and their army is quickly turning into a powerful internal security force. It is their political masters and "peace-makers" who have yet to grasp the significance of events and their implications. They choose, however, to absorb them into the obsolete paradigm of the "Israeli-Arab conflict".

Now that the Green Line has been re-established by Palestinian stone-throwers, and PLO chairman Yasser Arafat has uttered the "Latin" phrases of diplomatic jargon, the curtain rises on the final act of the Jewish-Arab drama. The perception is that radical change approaches and that this will be accomplished by the success of the now inevitable peace process.

This is a normative forecast not a realistic one. Following the 1987 uprising the internal Israeli-Palestinian conflict is even further from a solution than it was before. The feelings of hostility, fear and hatred have intensified rather than subsided. The dynamics of the internally generated conflict engender more violence and harsher enforcement. The *intifada* is itself the status quo.

Anyone who wishes to estimate the time span of what I term "status quo" should remember that the current "troubles" in Northern Ireland began in 1969 and that the Sharpeville massacre, which triggered the violent racial confrontation in South Africa, happened in 1960.

The eruption of the uprising in the territories, now in its third year, marks the final act in the process of the internalisation of the Israeli-Arab conflict and its transformation into overt, Jewish-Palestinian intercommunal strife. The essence of the new phase lies in the fact that the Palestinians have at last healed their internal schisms, overcome their lack of motivation and crossed the threshold of fear. For the first time in their history they have succeeded in mobilising the entire community in a sustained political struggle and in waging a controlled confrontation guided by realistic and not emotional considerations.

The leaders of the uprising have permitted no deviation from a quite rigid strategy that makes constant allowance for the population's capacity to endure hardship. This realistic strategy is manifested by limited commercial strikes, by allowing workers to continue their employment in Israeli enterprises, and, especially, in the ban on the use of firearms against Israelis. Twenty-one years of intimate acquaintance with the Israeli polity have taught the Palestinians the political and moral constraints that inhibit the Israelis' freedom of action and restrict their ability to employ their immeasurably vaster physical power. The Palestinians have turned these constraints to their own benefit.

It was the realism evinced by the leaders of the uprising that enabled them to persist with the revolt and reap the political fruits of the *intifada*. However, it was beyond them to formulate a positive political programme: that was the exclusive prerogative of PLO institutions.

However, the Arabs in the territories have broken with the past by refusing to allow the PLO to remain captive to its own illusions. They sent Yasser Arafat a clear message: the uprising must be translated into a realistic political programme. The illusion that Israel does not exist because the Palestinians are unable to accept its existence, or that it can be wished away, is a recipe for disaster. To go on ignoring the true power relationship would result in the loss of the little that can still be salvaged.

This was the juncture at which Arafat found his calling as a leader. Submitting to the constraints of reality, he sacrificed the maximalist formulations of the Palestinian revolution in favour of a political plan that recognised the facts of life, the cardinal component being acknowledgement of Israel's unassailable, permanent existence.

Realpolitik is not normally considered an unusual trait in a political leader, but the historic significance of the shift in the Palestinian posture at the PLO conference in Algiers and in Arafat's Geneva speech can be assessed only against the background of the traditional Palestinian national strategy.

In contrast to Zionism, which opted for a strategy of stages and pursued attainable objectives, the Palestinian movement refused to compromise on its ultimate objective, even if by doing so it could have made substantial short-term gains. The Palestinians always believed that no compromises should be made on matters of principle in a just cause. Had they reconciled themselves to reality and been willing to forgo the sanctity of the principle, they might already have been celebrating the 40th anniversary of the State of Palestine in half of the homeland. In 1988 Arafat grasped that subjective will cannot overcome objective reality and forced this realism on his movement. He announced that he recognised Israel's right to exist and renounced terrorism. The question is whether this new realism, which was not forthcoming when it was needed 40 years ago, will now permit the rapid attainment of the new and relatively modest goals of the Palestinian national movement.

However, the new Palestinian programme lacks a vital element: it refuses to accept the fact that, unlike in 1947, the Palestinian state cannot be established by "the justice of the nations". The price of refusing to accede to the Partition Resolution of 1947 at the time was that Israel in the meanwhile has become a vital independent actor with impressive manoeuvrability in the international arena. If in 1947 independence was achievable in an international forum, that same independence is now obtainable only from the Israelis, who occupy the entire land with overwhelming force.

But not even Arafat is capable of demonstrating this level of hard-headed realism. He continues to believe that international pressure and a political campaign aimed at third parties will eventually succeed in twisting Israel's arm. Essentially, he thinks that the United States is a replica of the British government of 1947/48 and that UN intervention to implement the Partition Resolution, intervention which the Palestinians rejected outright in 1947, will materialise now.

"Israel is not a principal side in the conflict," Khaled al-Hassan, Arafat's principal associate, told the Egyptian paper *Al-Mussawar* in January 1989, "for it is merely a corporation turned into a state, as has been said about General Motors. We are aiming at the Americans more than at the Israelis."

It is hard enough to admit, finally, that 1948 is irreversible and the Jewish state a *fait accompli*. It is unbearable to admit a catastrophic error that resulted in the loss of two generations and allowed the enemy an irreversible head-start of 40 years. Therefore, Israel must remain for the PLO an object, manipulable by external forces, and external diplomatic activity must be the means to achieve Palestinian national objectives.

It is fascinating to contemplate the reversal of roles that has occurred in Palestinian and Zionist perceptions. In 1947 mainstream Zionists felt that they had to salvage what was still salvageable. The aftermath of the Holocaust forced them to agree to the partition of Palestine without Western Galilee and without Jerusalem. The Palestinians, confident of their strength, refused to compromise with a minority they perceived as an alien element, doomed to extinction once the colonial power departed.

Four decades later things have come full circle. The present debates among the Palestinians are not substantially different from the debate in the Zionist movement in 1947 between revisionists, who believed in "Greater Israel" and rejected the partition plan, and mainstream Zionists who supported it. Now the Israelis in their turn are confident of their strength, believe in holding on to everything, and deny the existence of the other side as an independent constituency capable of resisting superior power.

The *intifada* and the strategy adopted by the PLO have forced the Israelis to confront the Palestinian collective directly. The cherished illusion of the "Jordanian option" and the perception of the conflict as an Israeli-Arab inter-state dispute has had to be abandoned. Yet most Israelis are incapable of coming to grips with the fact that they are confronted by an independent and autonomous constituency which claims symmetrical attachment to the land. For the Israelis to internalise this fact would mean forgoing the exclusivity of their moral and historical claim and redefining their national ethos. They could evade this so long as the Palestinians played childish games of terrorism and "armed struggle", allowing the Israelis to perceive the conflict in terms of physical survival.

The new Palestinian sophistication has caught Israelis by surprise, so long have they been accustomed to a rash, emotional and unrealistic adversary. The new situation calls for sophistication on the part of the Israelis, which is emerging.

The Palestinian challenge is being met by entrenchment behind the status quo and the floating of "peace plans", whose only test lies in their unacceptability, but which at the same time create an impression of movement.

However, a more serious attempt at sophistication is in the offing: a redefinition of the Palestinian problem. New Israeli plans call for free political elections in the occupied territories, army withdrawal from populated areas, and self-rule. Beyond the wish to drive a wedge between Palestinians in the territories and the PLO, there is an attempt to reduce the problem to the fate of the residents of the West Bank and Gaza: for the rest, the 1948 exodus is irreversible.

Neither PLO terrorism nor the mighty Palestinian state frighten the Israelis. What frightens them is the return *(al-Awdah)* of Palestinians, the principle even more than the actual return. A Palestinian state, just like the Jewish state, embodies the principle of the ingathering of the exiles. It is Zionism in reverse. One of the first laws to be enacted by that state would no doubt be the Law of (Palestinian) Return.

The return — a perpetual, unremitting endeavour — would be the *raison d'etre* of the state of Palestine as it is the fundamental credo of the state of Israel, no matter how many actually return. Palestinian national objectives, like Zionist objectives, would not be completely realised by independence and would produce messianic and chauvinistic impulses as powerful as any produced by the Zionist movement, and as uncontrollable.

Israelis may believe solemn Palestinian vows to forgo claims for land in Israel proper, just as Palestinians may believe that the Zionists have given up their expansionist philosophy. Israelis of all political persuasions may sense that they must cope with the two million Palestinians who remained in the land, although that realisation is growing very slowly. To acknowledge, however, that Palestine is the homeland of the entire Palestinian people, almost five million strong, is insupportable.

Such is the macabre irony of history. It was the Zionist claim to Palestine as the homeland of the entire Jewish people and demands for unlimited Jewish immigration that incensed the Palestinians more than political demands for a Jewish national home. That role reversal is well understood by users of the intimate Israeli-Palestinian vocabulary.

The PLO reacts vehemently to any attempt at fragmenting the Palestinians, not merely because it threatens Palestinian unity and the PLO's exclusive leadership role. They understand the dangers inherent in Israeli sophistication: what if third parties, the United States in particular, view the new Israeli plans as satisfactory? After all, the occupied territories and the violence there, not the 1948 refugees, are perceived as the core problems. It is therefore imperative to nip in the bud all attempts at beginning a dialogue between Israelis and an "alternative leadership" (in PLO parlance) in the territories.

Local leaders are not tempted by Israeli overtures because they would become traitors to the cause, and rightly so. No Palestinian can forgo the principle of the unity of two-and-a-half million Palestinian people in the homeland and an equal number of exiles in the diaspora and disown total allegiance to their recognised and united leadership, the PLO. Yet local leaders must contend with the reality of Israeli coercion, the appalling price of the *intifada* and the limits of their people's endurance. If the PLO bid for a comprehensive solution does not succeed soon, there will be no choice but to deal directly with the Israelis, thus threatening the united front.

Perceptive Palestinians are seized with the terrible apprehension that they will awake to find they have driven themselves into a blind alley; that the diplomatic world they entered after so much travail is nothing but a prestigious debating club in which points can be scored but nothing concrete achieved; that American recognition is very far from being a Palestinian Balfour Declaration; that, after raising all the flags on all the Palestinian embassies around the world, they will find that Israel holds higher cards in the game of nations and that a Palestinian state will not be achieved by the justice of the nations.

Perceptive Israelis, too, are worried about the morning after that would follow success in deflecting the PLO political campaign. Winning the diplomatic game would not change the realities on the ground. In the shared homeland, from the Jordan River to the sea, there exist two cohesive national communities engaged in a total civil war that by now has become a way of life, an endemic and organic conditon.

Israeli success in breaking up Palestinian unity would force them to confront a leadership even more formidable than the ageing PLO: local leaders hardened by Israeli harassment, closely acquainted with the Israeli political system, and capable of manipulating its internal weaknesses and divisions. Those leaders would relentlessly continue the internal struggle and the Israelis would find it more devastating than conventional wars.

Yasser Arafat would like to believe that he is a Palestinian Ben-Gurion. But the "real" Palestinian Ben-Gurion now lives in a refugee camp near Nablus. He's a 22-year-old youngster perhaps trying to organise a new shock force after the arrest by Israelis of members of previous units. He is a realist to the marrow and he knows what Ben-Gurion also knew in his twenties: national independence is attained by stubbornly building economic, social, educational and military communal power and not through declarations and speeches in exclusive international clubs.

The task of the Palestinian Ben-Gurion is easier than that of the Zionist hero: Ben-Gurion had to build a society from scratch, whereas the Palestinian need only consolidate a community already living on the land. But the Palestinian leader faces a greater external challenge: Israelis are not a colonial power but a powerful, indigenous people deeply rooted in the land. No force in the world can suppress the national aspirations of a cohesive national community. But there is no law of nature guaranteeing that those aspirations must be attained. The Palestinians had an opportunity in 1947 and they missed it. Paradoxically, the fact that they are now ruled by the Israelis offers them a second chance. The *de facto* bi-national entity created by continued Israeli occupation and the socio-economic-cultural interaction between Israelis and Palestinians (and confrontation is a powerful form of interaction) have brought about an internally generated conflict, waged by two intimate enemies.

With time, the reality of daily friction distances those who experience the agony from even involved outsiders, and draws the Israelis and Palestinians together. Such conflicts have two basic theoretical solutions: partition or power-sharing, or a combination of both. At this point in the discourse conflict-resolvers begin to discuss draft agreements. Such devices, however, require that the parties directly involved feel compelled to make use of them, which is not the case when the protagonists consider conflict resolution devices only through the prisms of their respective gains and losses, not as means to resolve their differences amicably.

A new approach is clearly needed. Instead of agreeing on what is achievable, one should start with an agreement on the opposite: what is impossible, unworkable or unachievable. That would lead to an open-ended process of problem-sharing.

The success of this process is not measured by approximation of results to a desired goal but by the degree of agreement, no matter how trivial the issue might seem. This process-oriented rather than solution-oriented approach requires a careful formulation of gradual steps, concrete and meaningful measures designed specifically not to contradict the diametrically opposed ultimate objectives of both sides.

Palestinian realism and Israeli realisation that they confront an unyielding Palestinian community may turn out to be the first steps on a very long road to reconciliation. They themselves must make the journey and there are no short cuts. While the international community can play a role in containing the dispute so that it does not spill over and threaten international stability, external forces cannot replace the communities engaged in the struggle. They must themselves realise that continued strife will bring a plague on both their houses, that nobody emerges triumphant from a communal strife and that one community cannot thrive forever on the misery of another.

7 A Palestinian view of the occupied territories

SARI NUSSEIBEH

During the late 1960s the Palestine Liberation Organisation (PLO) formulated its policy calling for a democratic state. This strategy implied that the PLO recognised the Jewish people although it did not recognise the Jewish state. It called for the dismantling of the Zionist entity or institutions and their replacement by institutions which were democratic and secular. There were problems in that call. From an Israeli point of view it was unacceptable because it meant the de-Judeaisation of Israel and an end to Zionism as an institutional form of political expression. This remained the creed of the Palestinian national movement until it was replaced by the two-states solution, which was finally accepted in Algiers in 1988 when the Palestine National Council (PNC) announced its Declaration of Independence. This declaration constitutes a major change in Palestinian strategy. The goal was no longer a democratic secular state in all of Palestine but a state for the Palestinians alongside Israel.

The Palestinian struggle is not a civil rights struggle. It has always been a national struggle, and the movement that represents it is a national movement. The majority of the Palestinian population who have created and been active in the Palestinian resistance are people who have been nationally and territorially dispossessed. Many are living in exile. Their immediate concern is not civil rights but the more primary need to return to their original homes and to re-establish a political state.

The Palestinian conception of the Israelis has been reformulated. In the earlier phase the Palestinians did not recognise either the Jewish entity as a state or the Jewish people as an entity. In the second phase, the Palestinians came to acknowledge the Jews in Israel/Palestine as an entity, although they still opposed their state. During this phase, they proposed a democratic, secular state as a final solution. Finally, in the third phase, Palestinians have come to acknowledge the people, namely the Jews, as well as their desired form of political expression, namely the state of Israel. Now the PLO has adopted the new strategy of calling for two states living peacefully alongside one another. The PLO is engaging itself in a political process which could lead to the creation of a Palestinian state in only part of Palestine. Obviously Israel and the Palestinian state would have to recognise one another.

The final push that made this change complete is the *intifada*, which from its outset has been a popular movement with a national ideology calling for nothing more than liberation and independence. These are the only two slogans that have been used in the entire literature of the *intifada*. This emerging sense of realism of the people under occupation has given leaders such as Arafat, who were already trying to lead the national movement in this direction, the political clout necessary to push the new strategy through the Palestinian National Council. The Palestinian strategy is now a clear-cut, two-state solution.

This does not mean that it is impossible for the strategy to change over the next 10 to 20 years. If a Palestinian state has not been established a decade from now and Israel continues to exhibit intransigence in the face of Palestinian peace offers, Palestinians may stand up and say: "Let us change our strategy to one of integration. Let Palestine and Israel merge into one state, and let our struggle become a civil rights struggle." At present, however, the people's political ideology and consciousness is different. People want to be freed from the Israeli system.

Palestinians are committed to making the two-state solution work. This means a fully-fledged, independent Palestinian state with East Jerusalem as its capital. Obviously there is a difference between principle and practice. In practice a fully independent state is not possible because the actual lifeline between the two parts of the state, the West Bank and Gaza, will depend on Israel's acceptance of this. This is the first of a series of dependencies and one can go further down the line. In every issue and every detail, in every aspect of the Palestinian state, there will be dependence and inter-dependence. One thinks of water supplies, in particular, which do not respect borders; or of establishing a Palestinian radio station where Israel's continued favourable disposition is the only guarantee of a particular wavelength. In reality, all our activities will hinge on Israeli co-operation.

The principle of sovereignty cannot be put into question, but when it comes to details and practical reality, everybody realises that there will have to be all kinds of interdependencies between such a Palestinian state and Israel on the one hand, and Jordan on the other; perhaps also Lebanon and Syria.

I do not believe the Israeli occupation of the West Bank and Gaza is irreversible. The major problem with Benvenisti's analysis is the fact that it is merely hypothetical and does not take into account the fact that a major event can fundamentally undermine the premise on which the deductions are made. The *intifada* has been such an event.

On an important level the *intifada* has been an uprising in the people's consciousness. In the twenty years preceding the *intifada* there was no collective consciousness of the fact that the Palestinian community was allowing itself to become integrated into Israeli society. The *intifada* was a revolution in consciousness in the sense that people became aware that this was happening. They also became aware that if they wanted to translate their nationalist dreams into reality, they would have to act. The first step was to begin a process of disengagement from the Israeli system.

The *intifada* has also created change on the international level, concentrating and intensifying international involvement in the peace process in the Middle East. This means that new opportunities have been created.

On another level, there is the impact of developments elsewhere in the world. Change in South Africa, with the government unbanning and talking to the African National Congress (ANC), may shock people in Israel into realising that one cannot ignore an enemy and that one must finally come to terms with it and talk to it. In the Palestinians' minds the ANC has always been identified with the PLO, and white South Africa with Israel. Many more Israelis will come to see the inevitability of talking to the PLO. There has been an increase in the number of people within the Israeli community calling for dialogue with the PLO.

Lastly, there is the question of Israeli security fears. Given Israel's military strength, I think these fears are exaggerated. But one has to distinguish between rational fears and irrational fears. Rational fears can be addressed and defused through mutually acceptable arrangements. Irrational fears can lead to untenable positions. For example, Israel could argue that it needed to seize the mountain ranges in Jordan, which are much higher than the West Bank hills, in order to ensure its security. Then having built settlements on those mountain ranges, it could argue it needed to occupy further territory to ensure the security of the settlers. Security issues can thus be used as a cover for territorial greed. Conversely, it is easy to conceive of a testable and workable arrangement which would involve a Palestinian state and a whole array of early monitoring systems on the ground and in space.

A Palestinian state could not establish an army that would have any kind of capacity to defeat Israel or even to defend the Palestinians from Israel. The Palestinians are going to be the party that will be constantly under threat from Israeli re-invasion. Therefore the Palestinians will have to be constantly alert in order to avoid provoking Israel or giving her an excuse to revert to the present kind of situation. In fact, it will be very advantageous and attractive from the point of view of both the Palestinians and the Israelis if a two-state solution is set up which is based on a system of co-operation and mutual interdependence.

If we agree with the Israelis on the principle of an independent Palestinian state alongside a secure Israel, and set a time-scale for the establishment of our independence, then almost anything can be agreed to as interim phases. If, however, one says "let us not talk about the final stage, let us just talk about the interim", then one will find Palestinians extremely suspicious. The reason is that they are afraid that any plan focusing only on interim stages and excluding an agreement on the final outcome will be used, not as a step towards liberation, but a step towards further naturalisation of the Israeli occupation of the West Bank and Gaza and the subjugation of the Palestinian people.

8 Economic growth and political absolutisms

NORMAN GIBSON

The issue I have been asked to address is: "Will economic growth and the removal of overt socio-economic discrimination lead to a significant tempering of absolutisms in Northern Ireland?" I begin with some manifestations of absolutism, briefly review some aspects of socio-economic discrimination and attempts to remove them, and then consider if these attempts and economic growth will temper the absolutisms.

Absolutisms

At the core of the communal divide is a deeply felt and pervasive sense of difference — two peoples as it were, self-consciously distinctive with different identities and cultural traditions. The sense of difference sustained and passed from generation to generation is often expressed in terms of "them and us", "the other side", Catholic and Protestant, Nationalist and Unionist, Republican and Loyalist, Irish and British. Political absolutism is conveyed by such slogans as "Brits out" on the extreme Nationalist/Republican side; "what we have we hold" and "not an inch" on the extreme Unionist/Loyalist side. One wants nothing but a united Ireland — constitutionally a unified state — and the other will not have it at any price — a collision of absolutisms; compromise is anathema, a betrayal, a form of treason.

Religious absolutism is for the most part more subtle but none the less real. At the centre of the Catholic/Protestant divide seem to be strongly opposed and apparently irreconcilable conceptions of religious truth and authority. These absolutisms are for the most part not openly and widely discussed in theological terms but manifest themselves in various ways. The most significant is perhaps education. Notwithstanding a growing interest in integrated education, recently encouraged by the Northern Ireland government, there are, with some important exceptions, essentially separate school systems, one for Catholics and the other for Protestants, reflecting by and large the preferences of Catholics and Protestants and their respective churches. It is hard to avoid the conclusion that this separateness reinforces and helps to sustain the divisions in the society, despite growing attempts to introduce into the curriculum such areas of study as education for mutual understanding and a more detached approach to the study of Irish history.

Religious absolutism also manifests itself in attitudes to so-called mixed or inter-church marriages. A recent study found that some 98 per cent of married Protestants and 95 per cent of married Catholics were married to someone of their own persuasion.[1] Of particular importance in marriages between Catholics and Protestants is the condition that "the Catholic party is to make a sincere promise to do all in his or her power in order that all the children be baptised and brought up in the Catholic Church".[2] This is perceived as oppressive by Protestants and ultimately a threat to their survival and identity.

Cultural absolutism or at least distinctiveness is also a feature of the situation, though perhaps its significance can be exaggerated as there is necessarily a shared or common culture in many aspects of life. Irish language, literature, folklore, music and sport are primarily the preserve of the Catholic/Nationalist tradition, whereas the Protestant/Unionist or Ulster Protestant tradition is, in the words of the distinguished Irish historian F S L Lyons, "the product conjointly of colonisation, of Calvinism and of industrialism". He goes on to describe the people of this tradition as combining "for much of their history a notable liberalism in public life with a strict severity of private discipline . . . a serious people, caring for education, intent upon self-improvement and material progress, yet deeply marked by their religion".[3]

In speaking of absolutisms there is a danger of distortion and oversimplification. The society is not rigidly dichotomised into two clearcut monolithic groups and the so-called absolutisms show some signs of attenuation. Despite this, however, the people remain deeply and selfconsciously divided and mutually distrustful, with for the most part each group lacking confidence in the other.

Absolutism in belief and conviction seems to lead frequently, some might say inexorably, to the development of institutions and structures which practise and perpetuate exclusiveness, often associated with illiberality and oppressiveness to outsiders, especially by majorities towards minorities or by one group towards another, large or small, in a position of dominant power. In some instances how this power is exercised may reflect a deep-seated insecurity, particularly if a group is actually or potentially in a minority and feels its identity and way of life under threat. The illiberality and oppressiveness may take many forms, not least social and economic discrimination. I turn to evidence of this in Northern Ireland.

Socio-economic discrimination

From the foundation of Northern Ireland under The Government of Ireland Act of 1920 many Catholics and Nationalists felt themselves to be second-class citizens and to have been involuntarily incorporated into the state as a minority — about one-third of the then total population of some 1,25 million. The sense of alienation and grievance persisted and eventually the Cameron Commission, appointed following the outbreak of violence and disturbances in 1969, found that much of the grievance related to allegations of discrimination in "housing and employment".[4] Further important issues were the restriction of the local government franchise to ratepayers and the gerrymandering of local government electoral boundaries to bring about and maintain Unionist control of local government in excess of their overall electoral share of the vote.

Each of these matters was subsequently addressed by legislative and other measures. The brief review which follows draws heavily on work of the Standing Advisory Commission on Human Rights (SACHR) but is not a comprehensive survey.[5] The Electoral Law Act (Northern Ireland) of 1969 extended the franchise for local council elections to all adults aged 18 and over. Other legislation "established an independent Local Government Boundaries Commission" to deal with the issue of gerrymandering. The Housing Executive Act (Northern Ireland) of 1971 "provided that all public authority house building" should be allocated "on the basis of an objective points system (and) become the responsibility of a central housing authority, the Northern Ireland Housing Executive". In the field of employment the Fair Employment (Northern Ireland) Act of 1976:

> made discrimination on religious and political grounds unlawful
> . . . established machinery for the promotion of equality of
> opportunity, (set up a Fair Employment Agency with responsi-
> bility) for receiving and investigating complaints of discrimina-
> tion and for conducting investigations into the extent of
> equality of opportunity. The Act also sought to promote

voluntary action by establishing a Declaration of Principle and Intent, to which employers and others were encouraged to subscribe. Those who signed the Declaration were certificated as equal opportunity employers.[6]

It is now widely accepted that grievances in relation to the local government franchise have disappeared. And the Policy Studies Institute report already referred to concluded "that there is a high level of confidence both among Protestants and among Catholics in the fairness of the system for allocating publicly owned rented accommodation".[7] Much the same was found to be true of the public sector house building programme.

But employment and unemployment remain areas of serious contention. Catholics persistently experience a much higher incidence of unemployment than Protestants. The evidence suggests that unemployment rates for Catholic males is some 2,5 times that for Protestants and has changed little since the early 1970s. This is illustrated by the data in Table 8.1. The differential between the unemployment rates for women is less pronounced than for men, though still large.

Table 8.1: Unemployment rates for Catholic and Protestants, 1971, 1981 and 1985-87 (per cent)[8]

	MEN			WOMEN		
	Catho-lics	Protes-tants	Ratio	Catho-lics	Protes-tants	Ratio
1971 Census	17,3	6,6	2,6	7,0	3,6	1,9
1981 Census	30,2	12,4	2,4	17,1	9,6	1,8
1985-87 CHS	36	14	2,6	15	9	1,7

The persistence and scale of these differentials are not by themselves sufficient to establish that in recruitment for the same jobs there is discrimination against Catholics and in favour of Protestants where each has similar skills, qualifications and abilities; other factors might account for the differentials but they certainly call for explanation.

The Standing Advisory Commission, relying on the report already referred to of the Policy Studies Institute on Employment and Unemployment, considered this matter at length.[9] It listed a series of factors adduced as possible or partial explanations, including the relatively higher concentration of Catholics in areas of greatest unemployment; the socio-economic classification of Catholics and Protestants, with the former relatively more concentrated in lower classes; differences in educational achievement and choice of subject specialisation; family size; the black economy and religion.

Each of these factors, and it was emphasised that they could not necessarily be considered as independent, was ostensibly found to be important, and some more than others. In interpreting the weight which should be attached to each it was also rightly emphasised that some, such as socio-economic class and the relative concentration of Catholics in the lower classes, might not be so much a partial explanation of unemployment differentials as an indication of the problem.

In attempting to take account of the inter-dependence of some of the various factors the Commission relied on a multiple regression analysis carried out by the Policy Studies Institute.[10] The broad conclusion was:

. . . that the predicted rate of unemployment is substantially higher among Catholics than among Protestants, no matter what are the other characteristics of the person under consideration. The results of the analysis conclusively demonstrate that after allowing for the factors included in the model there is still a large difference between Protestants and Catholics in the chance of being unemployed. It could be argued that factors not included in the model might explain some of the difference, but the model does include all of the factors that are known to be important determinants of unemployment.

At first sight this finding seems definitive and disturbing and the Commission concluded that "while all the listed factors, and particularly socio-economic states, are significantly related to unemployment, a man's religion is consistently shown to be a major determinant of his chance of being unemployed".

Multiple regression studies are notoriously difficult to interpret. The specification of the model or models and the accuracy of the sample data are critical to the meaningfulness of the exercise. In this case, dealing with such an extremely complex phenomenon as unemployment, the outcome of the interaction over time of both demand and supply factors in diverse labour markets, it may be doubted if a single equation type specification adequately captures the influences at work.

More particularly the analysis ignores the possibility of "self-exclusion" from or avoidance of certain occupations. The most important are security occupations, since for Catholics to participate in these could literally put their lives and families at risk from paramilitary reprisals and for political or other reasons might not be congenial. Nevertheless, this omission is curious since the Commission in its report refers to "the reluctance of many Catholics to take up employment opportunities in the police and other security services" for the reasons just mentioned and indicates that these occupations "currently provide up to 30 000 jobs in Northern Ireland, amounting to more than 5 per cent of total employment". It goes on to state that greater equality in the distribution of these jobs would make a significant contribution to reducing the differentials in employment or unemployment".[11]

Indeed, if it is assumed that there is a 40/60 split between Catholic and Protestant in the economically active male population and the respective unemployment rates are 25 per cent and 10 per cent — a 2,5 times ratio — and if employment in security occupations was shared proportionately, without other offsetting changes, then the corresponding figures would be 20 per cent and 13,3 per cent, reducing the ratio to 1,5. (For higher levels of unemployment with the same initial ratio of 2,5 the reduction would be less.) Clearly, ignoring this factor in the multiple regression analysis calls for qualification of its general conclusion. It would, however, be going too far on the basis of the evidence so far available to infer that religious affiliation can be explained away as a determinant of a person's chance of being unemployed. But at the very least further careful study is required of what is an exceedingly complex and critically important issue.

The government has not felt able to await the outcome of such studies and in 1989 enacted a new Fair Employment (Northern Ireland) Act with the objective of ensuring equality of opportunity in employment. It established a Fair Employment Commission with stronger powers than its predecessor, The Fair Employment Agency, to promote fair employment practices and audit and supervise their operation. The Act requires employers to register with the Commission, monitor the denominational composition of their employees and review their recruitment, training and promotion practices at least once every three years. In implementing these reviews employers will have to observe a code of practice and decide on what affirmative action is "necessary to ensure fair participation for both communities, and to consider . . . the practicality of setting goals and timetables . . . the Bill contains a range of criminal offences and economic sanctions . . . In particular, all public sector contracts . . . and government grants can be withheld from employers convicted of default and on whom the commission has served a notice of disqualification".[12]

This Act goes substantially beyond the provisions of the Fair Employment (Northern Ireland) Act of 1976 but, it is claimed, avoids introducing "quotas" or "reverse" discrimination and retains "the merit principle at the point of selection . . . (and) . . . recruitment on the basis of ability and aptitude".[13] It remains to be seen how the apparently conflicting aims of having "goals and timetables" whilst rejecting "quotas" and "reverse discrimination" will be reconciled in practice. But what cannot be doubted is a determination by the government to remove "overt" discrimination in the field of employment.

The administration of the law and the activities of the security forces are two further areas which are frequently subject to criticism by Catholics and to a lesser extent by Protestants. A substantial minority of Catholics consider the courts treat persons of their denomination unfairly whilst amongst Protestants it is the lower working class and supporters of the Democratic Unionist Party who "tend to be most critical of the courts".[14] As regards the security forces — the police, army and Ulster Defence Regiment (UDR) — there is widespread confidence amongst Protestants that they treat Catholics and Protestants equally, but with a minority of Protestants not entirely sure that the UDR is fair to Catholics. "A majority of Catholics think that the police and the UDR are biased against them" but have more confidence in the fairness of the army.[15] These views were recorded in 1986 but there is little reason to believe they have changed substantially in the last few years.

The activities of the courts and the security forces are kept under regular review by the SACHR. Of particular concern is the trial of terrorist offences under the Northern Ireland (Emergency Provisions) Act of 1978 by one-judge courts without a jury. A majority of SACHR has strongly recommended the introduction of three-judge courts and is unanimous that in the context of paramilitarist violence and intimidation it is not feasible to have trial by jury for terrorist offences. The government has so far not accepted the recommendation about three-judge courts.

The SACHR also maintains a watching brief on the activities of the security forces, paying particular attention to complaints procedures. These have been developed and improved considerably over the years.

Economic growth

Group economic self-interest and the protection of privilege are often of enormous significance in socio-political conflicts, giving rise to dominant and subordinate groups, possibly with complex dynamic interrelationships between them. This might suggest that if the benefits of economic growth were more widely shared by such a society, and if the privileges and power of the dominant group were reduced, conflict might become less acute. If, however, the fundamental basis of the conflict is perceived, implicitly or otherwise, to be an irreconcilable collision of identities, then it may be doubted that economic growth or greater distributive justice, important though these may be, will resolve the underlying conflict.

Economic growth in Northern Ireland, as conventionally measured by gross domestic product (GDP), was more rapid throughout the 1960s and most of the 1970s than in the rest of the United Kingdom and has since broadly kept pace with the latter. In general, average standards of living would seem to have risen markedly over the last 30 years for both Catholics and Protestants. Even so, GDP per head in Northern Ireland remains some 25 per cent below the United Kingdom average and much lower than in any of its constituent countries or regions.

Information about the distribution of income (or wealth), particularly between Catholics and Protestants, is not readily available and it is necessary to rely on such indirect evidence as the proportions of each in broad socio-economic groups. But it is not possible to compare such groups strictly accurately over time; this needs to be kept in mind in assessing the data in Tables 8.2 and 8.3.

In his article "Religion and Occupational Class in Northern Ireland", Edmund A Aunger found that Protestants as a group were relatively more affluent, with 41 per cent in the two top groups in comparison with 31 per cent Catholics and correspondingly 15 per cent and 25 per cent in the unskilled/unemployed category.[16]

Table 8. 2: Religion and occupational class 1971, economically active men and women* (per cent)

Occupational Class	Catholic	Protestant	Ratio P to C
Professional/managerial	12	15	1,3
Lower grade non-manual	19	26	1,4
Skilled manual	17	19	1,1
Semi-skilled manual	27	25	0,9
Unskilled, unemployed	25	15	0,6

(*Note: Farmers, farm managers, market gardeners and members of the armed forces were excluded from the classification and the unemployed were all included as unskilled.)

Similar data for the mid-1980s, provided by the Continuous Household Survey and shown in Table 8.3, indicates that Protestants have 40
per cent in the top two groups and 7 per cent in the lowest, contrasting
with 26 per cent and 9 per cent respectively for Catholics. Thus the
general picture suggests that Protestants as a group are relatively better
off than Catholics. But it is surely suggestive that substantial proportions of Protestants and Catholics are distributed across all the socio-
economic groups. There seem to be no clear cut socio-economic distinctions between Protestants and Catholics and, to that extent, little in
the way of class-related privileges. This is not to say, however, that all
discrimination is absent.

Table 8.3: Occupation and socio-economic group; all persons aged 16
and over, 1985-87 (per cent)

Group	Catholic	Protestant	Ratio P to C
Professional/managerial	5	11	2,2
Other non-manual	21	29	1,4
Skilled manual	20	20	1,0
Semi-skilled manual	28	23	0,8
Unskilled manual	9	7	0,8

Tempering absolutisms

Intractable and powerfully persistent conflicts of identity and culture
and ethnic rivalries are a major feature of the modern world to be found
in most if not all political systems. These conflicts and rivalries
frequently seem to take on an absolutist character which often excludes
or limits compromise and constructive plurality, and evokes violence.
The conflict in Northern Ireland is typical.

Discrimination is often a major aspect of such conflicts, giving rise to
a sense of oppression and alienation in the subordinate group or groups.
Steps can be, and in Northern Ireland are being taken to remove various
forms of discrimination. The issue here is whether these steps have
tempered the underlying absolutisms or ameliorated the actual or
apparent irreconcilability of the conflict.

As already suggested, these absolutisms are grounded in deep-seated
belief systems, concepts of personal and social identity, which help to
shape and form the distinctiveness of individuals and groups. They do

not seem to yield easily to social and political measures, let alone oppression. In a sense they would not have an absolutist character if they did. There is little reason to believe that a fairer voting system for local government elections, more equitable housing practices, stronger attempts to achieve equality of opportunity in employment, improvements in the operation of the court system, more effective accountability of the security forces, faster economic growth and a more equitable distribution of income and wealth would, in any reasonable time period, significantly temper the underlying absolutist feelings and convictions.

If, as has been argued, these issues relate to the concept of identity, both individual and group distinctiveness, to the very core of personal and group existence, then it is hardly surprising that they cannot be traded for socio-economic betterment. Thus it is a false hope to assume that the removal of discrimination and similar measures will necessarily lead to a softening of conflicts of identity. "Give them jobs and housing and they'll live like Protestants", attributed to a distinguished (liberal) Unionist politician was doomed to disappointment.

It also follows that a dominant group holding on to political and economic power, even at the risk of serious political instability and violence and the corruption and destructiveness this brings, is not necessarily holding on to them just for the sake of having them but as a means to an end, where the end is perceived as the survival of a distinctive culture and identity. Practically every socio-economic and political situation then comes to be seen as a zero-sum game.[17] Negotiation and compromise seem to be impossible. Gains for Catholics/Nationalists are seen as losses by Protestants/Unionists and vice versa, or at any rate by the most extreme on each side. This stance may also help to explain why Protestant and Catholic workers, despite apparently similar economic and class interests, have great difficulty in pursuing them in unison and unequivocally.

The perennial fear of many Protestants/Unionists in Northern Ireland is that any loss of political and economic power would be a step towards a united Ireland in which they would be a permanent minority, less than 20 per cent of the population. They fear becoming second-class citizens in a state dominated by Catholics/Nationalists in which their civil and religious liberties and their survival as a distinctive people would be at risk. Equally there are Catholics/Nationalists who believe that their culture, identity and status will never be properly recognised within the confines of Northern Ireland as a part of the United Kingdom. The nature of the conflict in all its pathos and tragedy seems to be perennial, intractable and, in short, absolutist.

If any escape is to be found from this sad impasse then it would seem necessary to challenge the basis and power of the so-called absolutisms or, more positively, to seek to develop the concept of pluralism. In the context of Ireland it is the whole island which has to be considered and the challenge has to be addressed to both Catholic Nationalism and Protestant Unionism.

The issue of the Catholic/Nationalist and Protestant/Unionist collision of identities was addressed in 1983, at least in part, by a remarkable initiative of the four main and constitutional, nationalist parties in Ireland: Fianna Fail, Fine Gael, the Labour Party and the Social Democratic and Labour Party. They established the "New Ireland Forum" as a means of holding public "consultations on the manner in which lasting peace and stability could be achieved in a new Ireland through the democratic process and to report on possible new structures and processes through which this objective might be achieved".[18] Studies on economic and other aspects of "a new Ireland" were commissioned, submissions were invited from various groups and individuals, including spokesmen for the main churches. The Unionist Party declined to participate though individual Unionists did take part.

The Forum concluded that a necessary element in a new Ireland was that "the validity of both the Nationalist and Unionist identities in Ireland and the democratic rights of every citizen on this island must be accepted; both of these identities must have equally satisfactory, secure and durable, political, administrative and symbolic expression and protection".[19]

Of particular interest was the evidence presented to the Forum by the representatives of the Irish Episcopal Conference (the Irish Hierarchy). Dr Cahal Daly, Bishop of Down and Connor (located in Northern Ireland) and chief spokesman, stated that:

> The Catholic Church in Ireland totally rejects the concept of a confessional state. We have not sought and we do not seek a Catholic State for a Catholic people. . . . We are acutely conscious of the fears of the Northern Ireland Protestant community. We recognise their apprehensions that any political or constitutional or even demographic change in Northern Ireland would imperil their Protestant heritage. . . . We have repeatedly declared that we in no way seek to have the moral teaching of the Catholic Church become the criterion of constitutional change or to have the principles of Catholic faith enshrined in civil law. What we have claimed, and what we must claim, is the right to fulfil our pastoral duty . . . to alert the

consciences of Catholics to the moral consequences of any proposed piece of legislation and to the impact of that legislation on the moral quality of life in society, while leaving to the legislators and to the electorate their freedom to act in accordance with their consciences.[20]

These last statements, though clearly not intended, give little reassurance to Protestants/Unionists who almost certainly interpret them as ensuring Catholic dominance in any form of a united Ireland where Protestants would be less than 20 per cent of the population.

Indeed, one of the Fine Gael participants in the Forum, commenting on its activities and particularly on the discussion of Church/State relations stated:

> . . . the probing and analysis proved uncomfortable, and none more so than in the area of Church/State relations and the role of the various churches in the finding of a solution. The presentation of the Catholic Hierarchy was one of the dramatic high points of the Forum.
>
> The Bishops performed with skill and subtlety but it was an essentially defensive performance which left too many unanswered questions and confirmed the extent to which the Bishops were, in some areas at least, as partitionist as the Unionists and as unwilling to yield an inch where their own vital interests were concerned. In all of this, the Forum was as much about the southern State and its willingness (or for the most part, unwillingness) to face fundamental change as it was about finding a lasting solution.[21] (Parentheses in original.)

Protestants/Unionists have not emulated the example of the constitutional nationalist parties in holding a "forum" about a new Ireland. Perhaps they have not had the confidence or they believe that it would expose and weaken their position. Whatever the reasons, if Protestants/Unionists want to share the island of Ireland with their Catholic/Nationalist neighbours in peace and mutual and dignified acceptance, then sooner or later they will have to find the courage to spell out the terms and conditions on which they consider this to be possible. It is perhaps a pipe-dream that despite everything an Ireland might emerge which rejoices in the rich pluralism of its traditions and sustains close, imaginative and constructive links with Britain and the wider world, but at least now there is the extraordinary example of events in Eastern Europe and the Soviet Union.

ENDNOTES

1. David J Smith: "Equality and Inequality in Northern Ireland, Part 3, Perceptions and Views", London, Policy Studies Institute, 1987, p64.
2. *The Code of Canon Law*, London, Collins, 1983, Can 1125, p199.
3. F S L Lyons: *The Burden of our History*, The Queen's University of Belfast, 1979, pp18-19.
4. "Disturbances in Northern Ireland: Report of the Commission Appointed by the Governor of Northern Ireland", Cmd 532, Belfast, HMSO, 1969, Chairman, Lord Cameron, (paragraph 129).
5. See sections 2.5 and 2.6 of the document "Religious and Political Discrimination and Equality of Opportunity in Northern Ireland: Report on Fair Employment", Standing Advisory Commission on Human Rights (SACHR), Cm 237, London, HMSO, 1987.
6. Ibid (parentheses added).
7. David J Smith, op cit, p108.
8. David J Smith, op cit, "Part 1: Employment and Unemployment", Table 2.1, p40, Policy Studies Institute, 1987; and Continuous Household Survey (CHS), Policy, Planning and Research Unit Monitor, No 1/89, April, Department of Finance and Personnel, Northern Ireland.
9. David J Smith, op cit, pp24-35.
10. David J Smith, op cit.
11. David J Smith, op cit, para 3.45, p34.
12. See *Parliamentary Debates (Hansard)*, House of Lords, Official Report, Vol 509, No 110, 28 June 1989, cols 796-797.
13. Ibid, col 796.
14. David J Smith, op cit, "Part 3, Perceptions and Views", p46.
15. Ibid, p147.
16. Edmund A Aunger: "Religion and Occupational Class in Northern Ireland", *The Economic and Social Review*, Vol 7, No 1, October 1975, pp1-18.
17. Compare Michael MacDonald: *Children of Wrath: Political Violence in Northern Ireland*, Oxford, Polity Press/Basil Blackwell, 1986.
18. New Ireland Forum, Report, Dublin, 2 May 1984.
19. Ibid, p27.
20. Ibid, "Report of Proceedings", No 12, p2.
21. Maurice Manning: "Twenty Turbulent Years", *The Irish Times, Supplement*, 3 October 1988.

9 A view from the Bogside

DAVID APTER

Londonderry, Northern Ireland. — A bomb killed a teenager
and wounded eight other people on Sunday during a parade
attended by Irish Republican Army (IRA) supporters.

The group apologised for the attack today saying it was
intended to kill security forces monitoring the march. The IRA
said it regretted the death of the bystander, Charles Love, 16
years old, which it called a "freak accident".

The Royal Ulster Constabulary said the blast wounded four
police officers, a soldier and three civilians.

The march was to mark the 18th anniversary of the day when
British troops shot and killed 14 people during a Roman
Catholic rights march. The constabulary, which serves as the
provincial police, said the IRA claim of a "freak accident" was
"another example of their nauseating hypocrisy".

New York Times, January 29, 1990.

Introduction

Poor Charles Love, dead at 16, and a mistake at that. His name didn't
help him much. There is not much love in the Bogside between contending forces. The newspaper clip sums up a good deal more about
violence in Derry than might appear on the surface. So many of the
tragedies are errors. So much of the violence occurs in commemoration
of previous violence, each episode adding a fresh quota of dead and
wounded. And, of course, there is both the irony of the original
occasion, a "rights march", and the charge of hypocrisy. (One must
consider the source.)

For most of us violence is something distant, outside, away from daily
experience. An event may occur, of course. Anyone can become a victim

anywhere. But as with lightning, a "freak accident" is for the most part random. Violence, on the whole, intrudes only through the newspaper or spot television news. Not so in Londonderry. There, as in Northern Ireland more generally, violence is part of a sustained pattern. Freak accidents are commonplace. People have learned to live on intimate terms with death.

But precisely what does that mean? How does life go on when one is living with death? It is a question which has become unfortunately relevant in many parts of the world. How can people maintain their sanity and continue to function when violence intersects at every turn: house and hearth, family and school, church and state?

I will take a stab at answering these questions by means of a quick descent into a community in which violence is a continuing experience, involving children, family and friends. Its causes are complex. But beyond a certain point, it is violence itself which generates more violence. It develops its own meaning, language, discourse and structure. Certain spaces become its peculiar terrain, a simulacrum for principles which, no matter how parochial the movement, universalise its significance.

Violence and organisation

Given these circumstances violence becomes less a departure from the norms of ordinary life than a condition which generates its own objects. That is, violence creates its own ordering discourse. Composed of events, translated into myths and logic, overcoming and heroic tales, it is complete with ritual death and sacrifice. Subsequent events constitute a re-enactment introducing solemnity and awe, endowing everyday events with particular meanings of their own. It is the discourse which redefines the community.

There are, as we know, whole societies structured around violence, by means of patterned reciprocities and exchanges of reward, risk and danger. When prestige and prowess become valued and meaningful in terms of retribution and revenge, a condition is created in which what Bourdieu has called symbolic capital (in contrast to economic capital) can prosper.[1]

Such societies appear to us exotic, strange, bizarre, unmodern. Order is not for them a matter of peaceful negotiation, mediation and accommodation, but rather a case of tit-for-tat (often validated by a cosmology of space and time).[2] To most of us this is politics turned upside down; violence defining order, not the other way around, or to use Gluckman's odd phrase, "the peace in the feud".[3]

But there is something unwarranted in this use of "us" and "they", this sense of the "otherness" of violent communities. Characteristics like these are by no means limited to "primitive" communities. They commonly accompany the revival of primordial affiliations, for example, race, religion, language, ethnicity, nationality. They can occur where economic capital is lacking and symbolic capital becomes a plausible alternative. Combine economic grievance with primordial affiliations and the result is explosive, as we see today in so many parts of the world.

What is interesting about the violence in Northern Ireland is that it shows within a relatively small compass how a lack of economic capital combined with a build-up of discriminatory grievances translates into symbolic capital, composed of myth and logic, story and text and providing a rationality of its own, a sense of destiny.

Both Catholic and Protestant militant bands have well established rituals of violence.[4] Acts of killing represent the exchange of symbolic capital just as in a "savage" tribe, but with one important difference. The vast majority so engaged see themselves trapped, locked in, disturbed at the continual erosion of civility. They know that demands for a settlement can only be illusory given the sheer force and implacability of political fundamentalism.

For symbolic capital has engendered power, embodying it in redemptive yearning. So much desire is packed into the project of overcoming that each explosion, each episode magnifies the backlog of accumulated grievances.[5] Memory so reinforced becomes cumulative. Speech and narrative articulate a romantic history of suffering, sacrifice and martyrdom, the political consequence of which include conviction and obligation. In these broad terms, what goes on in Northern Ireland has its counterpart all over the world, from Baku to South Molucca, from Kashmir to Lebanon.

The violence that characterises Northern Ireland is by most reckoning sufficiently explicable in terms of the continued exercise of multiple forms of discrimination and injustice against the Catholic minority by representatives of the Protestant majorities, and the British authorities. But there are other ingredients: the long story of how the patrimony was lost — lands to the Protestants, power to Great Britain.

Within this story of decline and fall are short stories, uprisings, Sunday massacres, bloody Thursdays, Easter Rebellions. Composed of many episodes, epic in form, these not only reinforce grievances and claims but also provide authenticity and a pedigree to which a logic must answer. Indeed, the logic serves to make the need for violence self-evident and its exercise self-legitimating. In such circumstances, even if some of the causes were eliminated, violence would continue. Plausible solutions are difficult to find when what began as cause becomes consequence.[6]

The events of 1968 in Northern Ireland then simply become a fresh round in an old story of past discrimination, prejudice, neglect, inadequate representation, irresponsible government and insufficiently developed political institutions. This has intensified discrimination by means of fresh grievance, a further validation of prejudice, and an increased unwillingness to compromise in both Protestant and Catholic communities. There is no basis for agreement on the meaning of terms such as "adequate representation", "responsible government", or the "development of political institutions". In this sense the past cannot really be rectified. Indeed, the search for solutions easily provokes new occasions for violence.[7]

In Northern Ireland one sees how the universalism of IRA political claims is embedded in particular circumstances, with the latter negating in practice what the former might present as principle. The texture of social life is shot through with primordialisms, rooted in family based kinship networks which serve as underground (as well as above ground) building blocks for para-military organisations. Rituals of violence are intertwined with allusions to more conventional ingredients of the sacred, and with reference to violated bodies (as in Long Kesh), self desecration, sexual suppression. Thus, if one begins with explanations of causes that are entirely rationalistic and logical, the social as well as the conceptual consequences will require going beyond a rationalistic model to a more interpretative one, or what might be called a phenomenology of violence.

A perspective

It should be very clear that this is not a case study in the ordinary sense of the word. That is why I have referred to it as a descent into violence. In the Bogside, as one quickly discovers, virtually everything that can be said is controversial, the words used, the people themselves, even place names on the map — Ireland, Ulster, Londonderry, Derry, the Bogside. Ulster smacks of Loyalism and Unionism. Ireland is an incomplete country waiting for its six wayward northern counties. To call Londonderry by that name is to take the Loyalist side. To call it Derry puts one on the Republican side. As for the Bogside, the name itself signifies lowland, wetland, Bog Irish, a kind of slime (good people live on the heights, bad ones in the depths).

These misnomers telescope. Like nesting boxes, each adds its quota of misunderstandings and grievances. New ones pile on top of old ones. The entire structure is held together by narratives, stories told in people's houses, recounted on street corners and in supermarkets, trumpeted from pulpits: a recitation of outrages in the recounting of which specific events and martyred names take on the voice of the Catechism.

This chapter deals only with a tiny slice of this ongoing narrative, focusing on one family living in the working-class Catholic ghetto in Derry called the Bogside, where most people are out of work.

Of course this is an unconventional way of considering violence. Most studies go from the top down rather than the bottom up and from a more all-inclusive rather than a particular vantage point. Here one is less concerned with an objective analysis than with obtaining a sense of how it feels to live a life penetrated by episodes of violence; to see how what appears to be senseless becomes sensible and in terms which make the change comprehensible.

I can make no claims to truly accurate representation, but my intent is honourable. It is to treat violence in the Bogside as a social text, to try to describe what one has seen in terms of how people define what is significant, while applying an interpretative rather than a functional framework.[8]

To do this, however, it will first be necessary to locate the Bogside in its setting in order to provide some insight into its special character as a community.

Londonderry/Derry

Londonderry (to use its full English name) was once a prosperous coastal town with an active seaport. Today it is a backwater in the real sense of the term, left behind by larger deepwater ports elsewhere, and has become more and more remote from the main supply routes of industry and container shipping. It is a place without much of an economic purpose. While it has some prospects for development the general picture is by no means hopeful. In any case it cannot control its own future, which is bound up with the economic fate of Northern Ireland more generally and, of course, the impact of the European Community.

Londonderry is caught in a series of predicaments. Even if the larger political issues could somehow be sufficiently resolved, and even if a tentative peace descended on the area, given the political circumstances of the region as a whole the future would remain problematical. Hence, not even with the best will in the world could a few "responsibles" plan intelligently for a future when the only thing predictable with confidence is the continuation of the violence.

Nevertheless, possibilities do exist. Among the conditions favouring a revival are a good university and college, and a countryside which is not only beautiful but a stone's throw (if one may use that expression) from the Irish Republic and one of its most dramatically scenic national parks. Moreover, despite the bombed-out buildings, ghettos (Protestant as well as Catholic) and graffiti everywhere, the city remains an architectural gem. Within its old stone city walls, it boasts a number of elegant small 17th, 18th and 19th century buildings.[9]

On the top of the hill is the Mount, entered through the Memorial Arch. It is the site of British military headquarters. The Arch is at one end of the main street, manned by a British army outpost. Passing through the Arch and into the Mount itself one can see out over the entire Bogside. Three television cameras mounted on a tower monitor the area. The buildings are inside a huge chain-link cage erected against bombs. (Indeed, the entire British military headquarters is in a cage.) From it, and especially at night, British army or Royal Ulster Constabulary (RUC) foot patrols move down the flypass constructed by the authorities which leads into the Bogside.

Whoever traverses this winding main street, whether by foot or vehicle, is under observation from the television cameras trained on it from three sides, the fourth being the entrance to the Protestant ghetto. The old fortress is extremely picturesque. Its antique cannons still look out over the lowlands beneath it to the hill opposite where rows and rows of identical Victorian houses march up and down. Reminiscent of midlands factory towns in England, the Bogside is a far cry from the comfortable houses in the suburbs of Londonderry. There where the middle class lives, the struggle is muted, the lines more blurred. One is face to face with "class struggles" in the Bogside. Nothing there is muted.

On the same road but at the opposite end of the arch is the Guildhall, a medieval-looking Gothic revival structure built around 1910 and housing the town council and some civic offices. The council meets even though several of the councillors have narrowly escaped death from car bombs. One very pleasant Protestant politician remarked, plausibly enough, that after the narrow escape of his wife and child he found it difficult to be neutral on civic matters. What he neglected to say was that he had been a key figure in the Red Hand Society, a Protestant terrorist organisation.

In the Protestant ghetto, which is surrounded by twenty-foot wire mesh fences, the walls of buildings are decorated with insignia of the Orange Order, medieval shields of clans, armorial insignia, and the sign of the Red Hand. In the centre of the city, surrounded by pleasant shops, gentrified establishments and a first-class bookstore, there is a statue of a British World War I soldier lunging with his rifle at a prostrate figure on the ground. It comes as a surprise that it has never been blown up.

Even under present political conditions people remain remarkably congenial. One of the oddities of political violence is that even those most deeply engaged in it can be on all other counts delightful, warm-hearted, hospitable, normal people.

For those who live there, Londonderry (the town) is a microcosm of the entire struggle in Northern Ireland: Catholic, Protestant, Republicans and British. But Derry is different in that it is a town with a Catholic majority in a country where Catholics represent only one third of the population. The rest are Anglicans and Presbyterians, who consider themselves gentrymen, commercials and farmers — in contrast to the Irish rurals who are considered to be "peasants". There is, of course, a complex local class and affiliational structure.

Because Catholics represent a larger proportion of the Derry community than they do in Belfast or Northern Ireland as a whole, Bogsiders have shown a greater willingness to participate in party politics than other more beleaguered Catholic communities. There is widespread sympathy (if not support) for the Social Democratic and Labour Party (SDLP), which most favours a peaceful settlement, and which has complicated connections to Sinn Fein, the political arm of the IRA. Of course, as one might anticipate, a good many quiet supporters of the SDLP are also deeply involved with the Provisional IRA.

The Bogside

The Bogside was originally one street at the bottom of the hill surrounded by the stone city walls (purported to be the oldest in Europe). Today it is a community like any other but with this difference: a great many of its activities centre around violence. Diverse in its source, violence is both planned and sporadic. The commitment of those who perpetrate it fluctuates. Few have not been involved in some way. Few remain "terrorists" in the sense that one could apply that term to the Red Brigades in Italy, or the Red Army Faction in Germany (Baader-Meinhof). The IRA Volunteers regard themselves as part of a professional army, but perhaps because of its underground status, they are part-time soldiers. Highly organised and disciplined for certain purposes, for other purposes this army's existence is sporadic. In any case its activities are closely linked to daily life in the Bogside.

Those to whom the word terrorist is most often applied do not recognise themselves in it. If people use the term at all, it has a very different meaning from the usual pejorative one used by outsiders. Most regard terrorism, if not in the way Lenin or Trotsky did (that is, as irresponsible activity outside of the framework of a theory of proletarian violence) then as the ambiguous outcome or an unfortunate conse-quence of a war in which hit-and-run tactics rather than pitched battles are characteristic. In the Bogside people do not deny that terrorism happens but they claim to eschew it as a principle of the movement.

Long before the present round of violence began there was unemploy-ment. In the Bogside unemployment is not an economic or statistical abstraction. Not only does unemployment affect Catholics to a significantly greater extent than Protestants, Catholics believe that the cards for self-improvement are still stacked against them. In the Bogside perhaps three quarters of the males are out of work. A good proportion of these have never held a regular job. Marginality is thus a long-term and self-perpetuating condition.

Along with it goes a culture of pariahdom, as if some original Irish sin descended from father to son, the cumulative effect of which provides generational continuity to hurtful memories. It inspires conspiracy theories: of political and economic neglect, and by design rather than accident. For a good many in the Bogside, these experiences run so deep that few people see any immediate prospects for a democratic political solution.

In the Bogside people are in and out of one another's houses, paying little attention to proper hours for meals. It is a life of conviviality and social intercourse which hardly sustains good property management. In contrast, Protestant areas — even the ghettos — feel more prosperous. People are more concerned with property and the visible caretaking of it. Primness and a manicured quality express the pride of the community. What the Bogside cannot transcend is the dole as a way of life.

While they are declining in number, the more radical "Officials" of the Bogside IRA believe that the real problem could hardly be the inability of democratic institutions to rectify what is wrong. Rather they believe the wrongs themselves are generated by such institutions when they are coupled with capitalism. The Provisionals take a less radical political view. There are very few Officials in the Bogside and they are increasingly residual in the movement generally. For the Provisionals the IRA is only one part of the total way of life. Some people can be *of* it without being *in* it. Some hold deeply radical beliefs on some counts while remaining conservative on others, such as the role of the Church. Convoluted as the political picture is, it is also true that in the Bogside at least, life is less violent and more diverse in outlook than in many Catholic ghettos, and with many articulated differences of opinion. So much so that it is difficult to remember on a day-to-day basis just how beleaguered a community it really is.

So easy and familiar is its discourse that one tends to overlook the special meanings of colloquial terms. In conversation one can hear talk about how difficult it is to keep body and soul together, without realising that terms like "body" serve as a code for heroic deaths and dying, while the "soul" is something that marches on, either towards the bosom of a Church which claims it as its own, or (whether liberal or radical) towards some salvageable and secular moral privacy and discretion. Indeed, in whatever direction each soul chooses to march, one comes up against the unfinished business that exists between religious commitment and secularity. Not all the enemies are at the gates of the Bogside. Some would include the Church itself.

Of course the presence of armed Protestants in their own ghetto on the hill (or the British soldiers in their barracks) helps to reduce internal conflicts considerably. This does not make them any the less bitter, but it downgrades all other concerns in favour of the conflict with the British. Despite the general atmosphere of warmth and conviviality, it is difficult for an outsider to discern the real texture of division within so small and yet so complex and intimate a community.

In the Bogside, if place names and ordinary words are subject to considerable controversy, terms like terrorist and victim, not to speak of Provisional, Official and Volunteer are more blurred than outside it. They are difficult to separate concretely from husband and wife, son and daughter, and all the friendship/kinship associations which large families breed. In the Bogside the family is big in every sense of the word.

Because the family is so important, political organisation is social. Neighbourhood associations, church gatherings and the administration of household affairs slip over into more clandestine forms of activity. Collect money for any of the above and some of it is likely to go for IRA-related purposes. Intentions then are multiple, the categories of everyday reality have a common sense peculiar to the history and circumstances of the Bogside. It is not at all a tribal society, yet it has its own boundaries, and they are in part primordial.

We can refer to the Bogside as a counter-simulacrum, a city at the bottom reacting to the city on the hill. In rejecting the latter, it accepts the classical inversionary task — to make the last first, to turn the world upside down, as familiar a process in Irish history as it was in 17th and 18th century England, and having much the same roots. These have been kept alive in the small political greenhouse that is Northern Ireland.[10]

We call the Bogside a counter-simulacrum, not because it represents the nucleus for a viable political future, something to be recreated on a larger scale, but quite the reverse. It is a community created by negativity. Cut off from the mainstream, depressed, oppressed, drawn into its otherness, it unites around a discourse which submerges more tense divisions. It is in this sense both an artificial community (a product of discrimination) and a natural one; a nesting place, part sanctuary and part target.

At the same time it is a very practical working society. People go about ordinary business much as elsewhere. It differs from the ordinary perhaps in terms of its connections to violence, and also in the degree to which its members have attained exceptional sensitivity. Virtually every location is endowed with special meaning — this tavern, that storefront, this barricade, that house or crossroads. Every recounted event has its sign, every signifier its signified, a symposium of the latter constituting an internal code, a lexical system. One might call the Bogside a discourse community.

In the Bogside then, every act is loaded with symbolic density and is the topic of intense discussion, and the more so because Bogsiders have the extraordinary articulateness of survivors. They are highly experienced in the nuances of language, its musicality, its tone. People are adept at connecting the activities of each day of their lives to the larger complicity of Republican struggle. Everyone demonstrates some degree of shared commitment, even when struggling to assert his or her individuality. The tone and the language reflect ecclesiastical intimacy, complete with referents of previous political sacrifice. Martyrs are retrieved from oblivion and grafted onto the tree of saints.[11] Even the most ordinary pamphlets and posters include a poetry of yearning, ecclesiastical in tone, full of moral urgency.[12]

People

Take for example the particular family I have in mind. Living in a three-storey house at the end of a street (across from a lot behind which stands the Bishop's Palace), in order of political importance the family consists of mother (employed), father (unemployed with small pension), and nine of an original ten children (of whom one is in jail for ten to fifteen years, two are hunger strikers recently released from Maze prison, one is dead, one is in the United States). Everyone plays some role in the movement. No-one is quite sure what the role of the other is. Despite the intense intimacy there is a certain separateness. Some things cannot be said.

On a Sunday the father will buy several papers. People come and visit. Children straggle down to eat at all hours. The newspaper headlines make terms like Ireland, Dublin, County Sligo, North Armagh, London, Parliament, and the Stalker Report immediate in terms of a particular event, yet also abstract and far away. The house is a sanctuary, both from the rain outside and the occasional British patrol or marauding Protestant band. It has its own steaming reality which admits only family and a few old friends.

Great Britain, the Irish Republic or the United States are far away. Each corresponds to stylised images, as in posters, or as they appear in the headlines themselves, unidimensional. But they also penetrate the Bogside in terms of money, taxes and work. In every family there are members who have emigrated to the US or England. And Noraid which supplies money and arms is almost as much a visible reality as the army patrols, the police and the Royal Ulster Constabulary.

A black flag flies from the window of the house which by day belongs to the older generation and by night to the younger. Between daytime and nocturnal preoccupations all abstractions disappear. The British night patrols, faces blackened, arms at the ready, are observed by the young from behind the lace curtains or the corner of a building. Hit or miss, both parties are hunter and hunted, each the other's prey. Face to face with that one is at preconception zero. One needs to start all over again.

Everyone makes some kind of contribution to "the struggle". Some serve as Volunteers in the IRA, others support them financially. Still others protect people against internal conflicts that get out of hand. In the Bogside, because violence is so commonplace, producing order is more important than committing an incident. It is a place where mediation is the first response rather than the last.

A sense of bounded terrain extends from the inside of the Bogside to the outside. Walk up the Mount, pass through the gate, turn left into the Protestant ghetto, and if you are a Catholic from the Bogside, you are almost certainly in physical danger. Open your mouth and the accent will betray what community you derive from. Perilous boundaries, then, are not only streets but accents, clothes. The signalling system has clear rules. Everyone follows them. It is important to be recognised for what you are. It is equally important to be in the right not the wrong place.

This condition troubles more radical members of the IRA, especially those who seek a way out by regarding the struggle as one of class rather than culture, religion, or history. They have tried to reach out to the Protestant working class, urging them to join in a common front against Britain. Such efforts have had little effect. Most Protestant militants, especially those who are working class, have much too well defined views about the "Paddies", reinforced by centuries of hostility. No amount of Marxist political doctrine will induce them to change. It is not their mutual working-class interests which bind them into a singular solidarity, but the mutuality of the struggle between them which sustains their separatism.

The Official IRA has not only attempted (unsuccessfully) to deflect terrorism and turn it towards nationalism and class struggle against the British and the bourgeoisie, but it has attacked religion and the Church. The result is that they have, to some extent, become outsiders within the movement. In the Bogside one can see how unrealistic it is to expect to erode the influence of the Church. Everyone knows that without the Church the family itself would quickly disintegrate. Moreover, with the Paisleyites and Unionists waiting at the gates, to call for a more common and secular front is to whistle in the wind.

Nevertheless, most families in the Bogside are truly and genuinely working class, classically so, and in an almost 19th century way. But, despite prolonged and widespread unemployment they are not "lumpens". Take the husband of our particular Bogside family. He is small, stalwart, upstanding and proud. He fills the coal scuttle, builds the fires, and keeps them burning. But he plays a secondary role to his wife. Intense, articulate, politically engaged, she is both a classic mother and a classic militant; the one who, with little formal education, has become teacher, social worker and political activist. He has been a goods handler for British railways, and more latterly a postman, which gives him a small annuity. But for most of his life he, like most Bogside males, has been on the dole and, to add insult to injury, on the *British* dole.

A good many of the older men in the Bogside give one a sense of being left over from the past and worn out by a life of poverty. They may be radicals in some sense but very few have fire in their bellies. Indeed, most seem tired of struggle, whether class or political. Almost all the older people have seen activist days in the movement. They age quickly, however, having built up sufficient moral capital to live on it without being called into active participation all the time. Yet they are right there in an emergency, pulled along by the militancy of their children who are quick to remind them of their obligations to previous martyrdoms.

For all the changes that have been induced in the Bogside and despite television sets and other electronic paraphernalia, the overwhelming impression is that of a time warp, like walking into a faded brown photograph taken about 1910. There is the father in shirtsleeves and braces, standing very straight. The faces of the children are pasty, undernourished, white. The stamp of poverty is over everything. It is a space warp as well as a time warp, a photographic freezing of row on row of Victorian houses and of debris.

Of course such a picture has no more genuine reality than a photograph itself. Very few people are set pieces. Only a few die-hards are fixed in their positions. There is determination to continue the struggle. Within that there is also continuous discussion of alternative possibilities. What is lacking is freshness. Outrage, while it focuses the intelligence wonderfully, does not leave much play for a constructive imagination. Yet even this needs to be qualified. For one also finds remarkably gifted insights. There are a number of extremely articulate writers, artists, journalists and generally concerned citizens, who are neither prisoners of the situation, nor willing to accept the situation as it is defined by the younger generation, the most militant, and those most willing (in their black berets, face masks and weapons) to go the whole way.

While much of the ideology is as old as myth and current as the latest funeral, there is within the Bogside an enormous critical intelligence which is provincialised less by the scope of its concerns than by the slim prospects for constructive social action. The tragedy is that there are so few acceptable institutional outlets through which to act. Among the more interesting intellectuals are not only journalists and writers but also priests, some of whom (given the conservatism of the Church) are made to appear as radicals, even though their views seem quite moderate. For them the predicament is how to find a political solution in a state which everyone considers to be corrupt and whose institutions (courts, parliament, police and churches) do the opposite of what they were intended to do: "courts which dispense injustice, parliaments which work for the advantage of only a section of the people, police who attack rather than defend, churches which contribute not to human dignity but to human pain".[13]

Violence in the Bogside is commonly seen from this perspective, as a natural consequence of legal or institutional violence against people. Hence the mutual killings, the flying squads of vigilantes on both sides, the choreographies of grievances. The reinforcement street by street, neighbourhood by neighbourhood, of separate religious communities is intensified by ecclesiastical jurisdictions which follow the same fault lines, prevent mediating institutions and restrict the flow of ordinary commerce. So too each community is physically defined, bounded by barbed wire, fences and defensive outposts.

Events

At the entrance to the Bogside, on the one remaining wall of a destroyed building, one sees the always freshly painted sign in large black letters, "You are entering Free Derry". Behind it, now derelict and with windows smashed, is an almost deserted, modern, cement high-rise apartment house, all the more bleak for the original red, yellow and other bright colours in which the doors and windows were originally painted. Its walls are covered with defiant slogans. It was here that the first civil rights actions took place and petrol bombs were lobbed at the police.

On this road in June 1968 Eamonn McCann and his friends brought John Wilson's caravan (where he lived with his wife and children in a "mucky lane" in the Brandywell district) to Lecky Road, the main artery through the Bogside, parking it in the middle of the road, blocking the traffic. Mr Wilson had been told repeatedly by the authorities that there was no chance that the town housing authorities would provide him with a house. One of his children had tuberculosis. The caravan was "an oven in the summer, an icebox in the winter". It was the perfect icon for a protest against housing discrimination against Catholics.

> We distributed leaflets in the surrounding streets explaining that we intended to keep the caravan there for twenty-four hours as a protest against the Wilsons' living conditions and calling for support. We then phoned the police, the mayor and the newspapers, inviting each to come and see. The mayor did

not come. We expected that the police would try to arrest us, or at least to move the caravan to the side of the road. But they merely looked and left.

We stood guard on the caravan all night, equipped with a loudhailer with which we intended to try to rouse the district if the police made a move. But nothing happened. On the Sunday we hauled the caravan back to its original parking place. We had about two hundred supporters with us on the return journey. Reports of the incident were carried with some prominence in the Irish Sunday newspapers and on local radio and TV. We announced that next weekend we would repeat the performance, this time for forty-eight hours. During the next week we were visited by policemen who explained, almost apologetically, that if we went through with this they would "have to take action", which greatly encouraged us. On the Wilson issue we now knew that if we were arrested we would have strong support in the area. The facts had been given wide publicity and no one could deny that a great injustice was being done. Few, therefore, could openly oppose the vigorous protest against it. If we could force the police to act against us we would be certain of an upsurge of sympthy which would further weaken the Nationalist grip on the area.

We lugged the caravan on to the road again the following Saturday and waited up two nights for the police onslaught. But again nothing happened. We dragged it back to its laneway and resolved next week to take it into the city centre. Before the week was out the Wilsons had been guaranteed a house and ten of us had been summoned to appear in court for contravening the Road Traffic Act (N.I), 1951. It was a perfect ending.[14]

It seemed a perfect beginning, not to the violence which subsequently occurred but to a rather modest form of extra-institutional protest.[15] Instead it triggered an escalating pattern of violence which has not only wracked Ulster but Britain as well. The old loyalties are continually reaffirmed. The Protestant Unionist militants proclaim their loyalty to the Queen while they condemn British democracy and Parliament. They believe they have been let down and betrayed by successive British governments and Secretaries of State for Northern Ireland. Equally, in the Bogside where the Catholics have their "republic", people believe that the present political incumbents of the Irish Republic have utterly betrayed them, making common cause with the British against the IRA, assisting and co-operating with the British authorities. They believe the

Irish Republic has signed a demeaning agreement whose effect is to freeze the political situation. They feel rejected, for they know that without the Republic there is no hope for the Catholic cause.[16]

The external view then is bleak. Yet within Bogside there remains both intensity and vitality. Murals on outside walls juxtapose Che Guevara, IRA Provisionals and the guns and berets of the Black Panthers. These and other classic revolutionary symbols conspire to saturate space and terrain with social and political reminders of the late sixties. Politically they also constitute alternative icons, the displacement of the religious iconography, a decentreing of the Bishop's palace. In the Bogside, if the sacred on occasion becomes violent, such violence is itself sacred.

Social life

But one does not want to exaggerate. In the Bogside things are still in their places. It is a community, not a war zone. Family life, though not normal, keeps up appearances. The family remains the crucial unit of responsibility. If more often than not the mother rather than the father is at the centre of it, it works, unlike so many ghetto counterparts elsewhere. Families are still large, the number of children ranging on the whole from five to fifteen. Almost all have had more than one member in jail. Everyone pools social security and unemployment benefits. Among those who pass the age of 11, only a tiny minority go on to college or university. Many of those who do so are disbarred from employment because they have a prison record.

In winter the atmosphere can be stifling. Coal dust settles on everything, particularly on the wet laundry. There is a superabundance of people in and out of houses and an overwhelming visibility of daily life, of caring for needs by primitive means. Eating involves many people at all hours, cooking on a coal or woodburning kitchen stove, hauling coal in scuttles up several flights of stairs. Clothes are hung to dry everywhere, there is an abundance of whiskey and beer bottles, unwashed glasses, and the inevitable smell of urine from overworked toilet facilities. Too many people and too few facilities lead to more or less haphazard cleaning.

Yet in many homes parlours or sitting-rooms sparkle, even though they are packed with overstuffed sofas, big armchairs and huge colour television sets at centre stage. Trinkets abound: mementoes of trips, photographs, the inevitable portraits of Christ and, of course, the crucifix. Everyone has scrapbooks, objects, letters and poems involving prison and the struggle.

A few drinks, and the elders will indulge their storytelling proclivities, dwelling on origins of the struggle and tracing it back to the Anglo-Norman invasion. For younger people, journalists, writers, intellectuals, artists and others returned from years in the United States or work in Britain, the struggle is defined by events near the Bogside itself, the protest march of 1968 which we have just described. They describe with humour the roadblock, the old man and his large family. But they have less humour over present circumstances — the miserable housing conditions of Catholic families whose plight is largely ignored by the predominantly Protestant housing authorities.

There is conviviality and drinking. Evenings are for friends and families to congregate, argue, discuss. In such personalised communities the blowing up of pubs and other meeting places is violation of private, not public space. For they are meeting grounds for intersecting networks of families. Each individual death is magnified, intensified, remembered.

By the same token, the family is brought into the violence not only by the acts each member commits but because each house is a possible IRA defence post. Hence British patrols enter houses, rip up the floors and pull out cupboards in search of guns. Violations of law are commonplace. How many captured Provisionals have been bundled into Castlereagh and Long Kesh without council and subjected to brutal treatment?

No wonder Long Kesh (better known as Maze prison) became for a time the counter-simulacrum within the counter-simulacrum, when the hunger strikers went naked, spread their faeces on their bodies and on the walls of cells, and lived in filth, mortifying the flesh like the early Christian anchorites. Undertaken to force attention on what they considered to be their military struggle, this episode ended with the martyrdom of Bobby Sands, while the others, having lost teeth or hair, survived as permanently damaged human beings yet serving as models for the young, symbols of total commitment. In prison they learned Gaelic, wrote poetry and history textbooks for instructing children on their struggle.[17] No doubt few people actually read Bobby Sands' *Skylark Sing Your Lonely Song*, but the title is enough, under the circumstances, to invoke the sense of exceptionalism in the movement.[18]

What prison provides is laceration, wounds, sacrifice, mortification of the flesh, prisoners retrieving the past martyrs of nationalism. The retrieval is better than the projections. One can speak of displacement, dispossession, the need to take possession through violence, and to do so in every sense as free citizens. But the actualities of the struggle are not as promising as the language implies. The reality is written on pathetic and pasty young-old faces of people whose lives are aimless except for the violence in which they participate.

Everyday life within the Bogside includes among its uncertainties the anticipation that anyone in one's own family might be picked up by the police, rightly or wrongly. Indeed, virtually everyone has committed a punishable offence at some time. Less likely to be known is whether one's children are part of the Mafia, the protection rackets, the corruption which also flourishes in the name of the IRA.

The Bogside then is not only a place. It is a state of mind. It represents a condition of negative otherness. If to Protestant eyes it is sweaty, fecund, noisy, drunk, lazy, violent, dirty, sloppy, careless about property and undisciplined — to the more sympathetic observer it is easygoing, fraternal, funny, tragic, warm, affectionate, in short, everything the Protestant communities are not.

There is a sense in which those who live in the Bogside accept both sets of judgments, expressing them as so many pariah peoples do in jokes told at their own expense. They do not so much reject such attacks as claim that the conditions are caused by discrimination. Violence is necessary not only to get rid of the original conditions, but to void the stigma of the poverty persona. So if there is considerable self-hatred within the Bogside, it is ironic, bitter, funny. And a good deal of the humour is about the menace of "boundary crossing", like the man walking alone at night in Derry who is grabbed from behind, a pistol pressed against his head. "Don't shoot," he yells, "I'm a Jew." "What kind of Jew?" comes the response, "Protestant or Catholic?"

The Church

Standing both above and inside the violence, as if it could mediate its morality, is the Church. Its highly ambiguous role, for and against, summarises the predicaments the Bogside confronts as a community. The Church, for all its ancient wisdom and affiliations, is defined by its religious enemies — the different Protestant sects, Presbyterians, the English Anglicans and so on.

It is also the primary affiliation, the institutional locus of the community, and coequal to the family as the nurturing instrumentality. It is the father to those mothers whose husbands are unemployed. It provides not only spiritual sustenance but continuity in a world of risk and dangers. It penetrates all the ordinary events of everyone's life, and its reminders are everywhere. In the Bogside the Church is omnipresent even when it is disliked.

It is the Church which prevents the Bogside family from disintegrating as it does in so many other inner city communities. It governs in place of the state. It regulates the cycle of births, christenings, funerals and saints days, and it resists with considerable dignity the relatively fascist fulminations of the Reverend Ian Paisley, MP.

Among its most frequent events are political funerals. These constitute occasions for celebratory outrage. In the Bogside they take precedence over the weddings and christenings, the life-giving occasions at which the Church so expertly officiates. The funerals help keep the Church intimate with violence. Bereavement preserves familial, tribal, primordial feeling, even among socialists.

The Church is also the source of dissension. It contributed to the split in the IRA between the Officials (Marxists), and the Provisionals. But even those who support the Provisionals tend to have strong or ambiguous feelings about the rigidity of the Church on matters like divorce, abortion, contraception and so on. The more radical do not want the IRA to join with the Irish Republic because they fear the domination of the Church, regarding the Irish Republic as priest-ridden, still stuck on issues of abortion and divorce.

Most in the Bogside population, in the last analysis, would prefer a secular politics. They understand that democracy, pluralism, elections, laws and orders means separation of church and state. They are not theocrats. The problem is more a practical one than a matter of principle. What will be the political unit? How shall it be constituted? Hence if the Irish Republic appears to have abandoned their cause and they can expect little mercy from the Protestants, it is the institutions of the Church which define both the moral and the physical architecture of the Bogside.

Communion, parochial schools and disciplining priests are pervasive, and with the IRA such a part-time organisation, the Church has never quite lost its monopoly. It is severe. It emphasises sexual repression and discipline. But it offers a way out. For it also accepts that violence, like sin, is fundamental to the human condition. The Church, then, extends its grace to those who commit violence, even while deploring it. It is the crucial link between family, school and community. If its schools and parishes produce a primitive form of understanding and quite often savage reprisals for transgressions, the Church also meets real error with tolerance.

Composed of families rather than individuals the congregation is the building block of social networks. The Church is thus the defining institution of both Catholic universality and separateness. The congregation listens when, with respect to the political struggle, the Church like all great institutions speaks out of both sides of its mouth. In the Bogside the Church is also accepted in part because it is the embracing, universalising, all-encompassing institution which connects the congregation to all parts of the world. The more radical clergy link the struggle with anti-colonialism and anti-imperialism everywhere. But at the same time the Church is uniquely parochial and provincial. What goes on in the Bishops Palace at the edge of the ghetto is the object of bitter controversy. Accused of hypocrisy when it addresses in broadly humanitarian terms the need for peace and mediation, the Church is unyielding, parochial, utterly loyal, and intervenes in every factional dispute when it sees its interests threatened.

As a living and past presence, the Church articulates and regulates virtually all the boundaries of daily life, marriage, wedlock, births, communions, confessions. It mediates between purity and danger, and applies an efficacious ritual, sacramental rather than magical. Commemorative of past acts, it links and signifies the Irishness of the past with current sacrifices and deaths.[19] Violence then defines the task of the ministry as well as the family, the two remaining in this context more deeply intertwined than ever.

Politics

The notion of politics in the Bogside has less to do with parties (illegal ones like the Sinn Fein and its leader Gerry Adams, legal ones like the SDLP) than with local government and housing, allotments and pensions, petitions and demands.[20] It is also, of course, about war, terrorism, the Volunteers in their black face-masks, berets and uniforms.

Ideology is first a mythical story of the fall, of the loss of the patrimony to outsiders, the yeoman becoming the peasant, about pariahdom. Such narratives are both antique and contemporary. There is re-enactment in each violent episode and plenty of experience to validate myth. A struggle of such sheer duration provides enough dead martyrs to replenish the record. As we have said, these provide authenticity, pedigree and obligation for the movement. Dying violently is the final proof of the legitimacy of a cause which would be lost except for the commitment of those who continue it.[21]

Nationalism, religion, language, class, culture, history are thus present in events. This discourse reinforces separateness. It emphasises the differences between Bogsiders and the British, Protestants and Catholics. It also universalises the struggle in terms of the oppressed against the oppressors, colonials against an alien power. There is an involuted quality, a sense of being cut off from the rest of the world, of being perpetually misunderstood, even by friends (Irish Americans in particular). But there is also a sense that the Bogside is a particular form of a more universalised struggle.

The first reinforces the Bogside as its own community, its own reality. It intensifies interior networks. It closes the community to outsiders and defines betrayers. It blurs conventional distinctions between civilian and military, youth wings and more elderly supporters, volunteers and those not formally engaged. Political activities include running soup kitchens, bandage brigades, early warning systems, and providing safe houses. With such activities go a paraphernalia of clandestinity, secret passwords, caches of arms, and so on. Honour is secrecy. Truth is not telling. Virtually all organisations have two faces, the one above ground which looks out to the world, and the one below ground which faces inward, operates in its own way, and on its own time, clocks and schedules.

In people's houses, as we have suggested, the day belongs to the outside world and the more ordinary business of feeding, resting, working — a time for grown-ups. When the older ones are asleep, it is the time for all the comings and goings, the movements of weapons, the tactical actions.

One can see how misleading formal descriptions of the IRA really are. It is a movement. There is a party which represents it, headed by a brilliant young intellectual, Gerry Adams, the president of Sinn Fein. Whatever one can say about the IRA as a movement is bound to be misleading. Is it terrorist? Well, yes, but it is more than that. Is it of long duration? It claims a continuity back through the Irish Republic Brotherhood and before, to 1916 and the Uprising to its semi-mythical ancestral figure, Theobold Wolfe Tone.[22] Here we enter an even earlier realm of political folk memories of marauding Protestant vigilante bands of the 17th century: the White Boys, the Hearts of Oak, and the Peep O'Day Boys who not only engaged in terror, but left their signatures on burned houses or swinging corpses, the violent pedigree of the Orange Order.

The IRA has reconstituted itself again and again, and divided as well. Almost two decades ago when it divided into the Official IRA and the Provisionals, the latter claimed allegiance to the Thirty-two County

Irish Republic proclaimed on Easter 1916. But the real battle between the Officials and the Provisionals, as described by Sean MacStiofain, former chief of staff of the Provisional IRA, was that the Officials proclaimed their Marxism, their secularism, their belief in a common class enemy, and refused to accept the reciting of the rosary at Republican commemorations.

For the Provisionals the struggle over religion is not simply a matter of belief. The Catechism is essential for secular purposes. It grafts nationalism onto religious experience.

Conclusion

As in some primitive societies, violence in the Bogside has created its own structure. It has established reciprocities and networks of prestige and power. These include organisations like Sinn Fein, the political arm of the IRA, and the Provisionals. Rituals of killing generate symbolic capital, the latter embedded in family, networks of kinship, age grades, neighbourhoods, work units. It is impossible to destroy the movement by force. If the IRA is organised around communities, it is above all a family affair.

Indeed, the Bogside is more of a community *because* of the violence. Forced to look inward, it has established its own rules, constraints and requirements. It monitors the paying of political dues and subscriptions, the care of ex-Volunteers. As for militancy and demonstrations of loyalty, the older generations are willing hostages of their children and grandchildren, all of them held together by the condition of negative otherness defined by the outsider Protestants.

It is impossible for each younger generation not to take its place in that succession of foot soldiers which has paid its dues to pariahdom, added its contribution to martyrdom, and inculcated in parochial schools a transformed negative otherness by means of an image of military discipline and revolution. Suffering is an old story, an extension of the suffering Church into the narrative of the Irish struggle for freedom.

More difficult is the logic to project, the solution to be found. Even if a political solution could be found, there is too much unfinished business, an inner city community overwhelmed by poverty and unemployment, insufficiency of education, lack of motivation among those who need it most. There is also the inheritance of colonialism, the leftover habits and circumstances so common in ex-colonial territories. The imprint of a residual colonialism remains in ambiguities of self-worth, even the half-liking for those who have imposed an alien presence.

How much does the IRA speak for Catholics? It is probably an irrelevant question, as much perhaps as when the PLO speaks for Palestinians. But a more mediating political party like John Hume's Social Democratic and Labour Party, also speaks for Catholics. And as far as Gerry Adams and the Sinn Fein is concerned, the SDLP has no real solutions to the problem of minority status as long as the IRA is rejected by the Republic of Ireland.[23] Indeed, there are those who argue that the economic future of the Bogside, even given electoral domination by the two-thirds Protestant majority in a six-county mini-state, would be better over the long term with Britain in Europe than with the Republic.

But economic considerations, while important, will not prevail. In any case, hardened attitudes show few signs of softening. All sides are opposed to direct rule from London. Yet without the latter no political settlement is remotely plausible. Despite all these difficulties, there remains — and especially among the older generation — a residual belief (although not much faith) that political parties must at some future time and circumstance serve as a venue for a more normal politics. And, for all the hostility to Brits, there is something like a working-class sympathy extending to working-class soldiers. For all its categorical differences, the Bogside knows its English, knows the context from which they come, has family in England, and it is to England that they go for jobs much more than to America. If such ambivalences can be found in other Catholic communities, such as West Belfast, and among Protestants as well, it will in the end be the ambivalent who will settle the accounts.

As to Protestants, there is a sense that they are such a familiar cause of a conflict that has gone on for such a long time that one can cope with them despite sporadic raids by extremists. Even though there are few connections or relations between Protestants and Catholics, there remains a cheek by jowl quality to the relationship. People in the Bogside consider them less of an enemy than the British. They regard the extremists among them as the real problem. Indeed, one finds very little direct antagonism expressed against Protestants. Nor should this be put down to an exercise in futility. To judge by the Bogside, there is in Northern Ireland an asymmetry of prejudice. Protestants are more prejudiced than Catholics. Whether this would be the case if the situation were reversed politically and the Protestants became a minority within the Irish Republic, it is impossible to say.

One hears plenty of antagonism towards the Irish Republic for its co-operation with Britain, its hostility to the Sinn Fein, and its conservatism in matters of divorce and abortion. Nor is there much sympathy for parochial education, although one would not want to push this matter too far. We said that violence generates violence, even though people are confused and tired. Nevertheless the movement depends on violence to keep things as they are, for each fresh break renews commitments. If someone is picked up by the police it is an occasion for solidarity, reassurance, the expression of sympathy and support. It is an affirmation of family and community. There are few complaints. If someone is arrested in the course of (or as a result of) an action, no one asks whether he or she is guilty. (Guilt is a term used only by those hostile to the movement.)

For the police, the British army authorities or the RUC, guilt is a matter of getting someone to confess. Hence the niceties of the law are frequently honoured in the breach. The key question for both sides is whether someone will break in prison or be able to hold out. By the same token, if someone is caught, the trick is to put the police or the army on trial. Everyone is aware that there is sympathy for the cause in liberal sectors of the British public (as the Stalker inquiry suggested, and the public condemnation of the Special Air Services [SAS] when it gunned down IRA militants in Gibraltar in early 1988). For those killed by the IRA, there is no remorse.

Where will it all end? The Bogside presents a picture of continuing violence and no solutions. Certainly one cannot look to political parties to resolve predicaments, especially since the abolition of the Stormont provides them with insufficient local venues for power. Terror will no doubt go on, for it provides its own kind of discipline in the general community and in people's own lives. Yet virtually everyone deplores it and its consequences, and what it does to their children. So while there is a sense of tragedy everywhere, people have learned to accept it. Yet for all such militancy there are neither many revolutionaries nor conservatives in Bogside. Most subscribe to some kind of socialism, the outlines of which are blurred.

What this brief excursion into the Bogside suggests more generally is that when violence (and especially primordial revivalism) are linked to conditions of marginality, wider beliefs and principles coincide with concrete interests. Violence may then erode a community less than it prevents its disintegration. Creating a peculiar form of functional significance among functionally superfluous men and outraged women, it restructures social life around a politics of confrontation, using ancient institutions like the Church. The latter provides a language of suffering which enables people to sustain commitments. But it also suggests that no matter how much anger there is and how deeply felt those commitments, no one believes in them totally. The main cause for hope in Bogside is that people believe and do not believe, all at the same time.

ENDNOTES

1. See Pierre Bourdieu: *Outline of a Theory of Practice*, Cambridge, Cambridge University Press, 1977.
2. See Bourdieu, op. cit.
3. See Max Gluckman: *Custom and Conflict in Africa*, Glencoe, The Free Press, 1953.
4. Walter Burkert, Rene Girard and Jonathan Z Smith: *Violent Origins*, Stanford, Stanford University Press, 1987.
5. See Paul Ricoeur: *The Symbolism of Evil*, Boston, Beacon Press, 1967. See also Northrop Fry: *The Secular Scripture*, Cambridge, Harvard University Press, 1976.
6. See Alan Bairner: "The Battlefield of Ideas: The Legitimisation of Violence in Northern Ireland", *European Journal of Political Research*, No 14, 1986, pp633-49.
7. Andrew Boyd: *Holy War in Belfast*, Belfast, Pretani Press, 1987.
8. For some of the theoretical concepts employed in this analysis see D E Apter: *Rethinking Development*, Newbury Park, Ca., Sage Publications, 1987.
9. Some are being rehabilitated by a training programme for unemployed youth, headed by a remarkable Catholic social worker who was one of the leaders of the 1968 march which triggered the chain of bloody events which have followed ever since. He and his co-workers have succeeded in rebuilding and rehabilitating what others have burned down (and in the process converted a number of potential burners into builders).
10. See, for example, David Underdown: *Revel, Riot and Rebellion*, New Haven, Yale University Press, 1987. See also Christopher Hill: *The World Turned Upside Down*, London, Penguin Books, 1972, *passim*.
11. IRA calendars list the dead by their various brigades, or by whether they are H-Block martyrs, and soon Saints' days blend with commemorations like Sunday, January 31st (Bloody Sunday).
12. For example, in an IRA calendar next to a photograph of masked IRA snipers in a trench one finds the verse of Padraig Pearse: "And now I speak, being full of vision;/ I speak to my people, and I speak in my people's name to the masters of my people,/ I say to my people that they are holy, that they are august, despite their chains,/That they are greater than those that hold them, and stronger and purer,/That they have but need of courage, and to call on the name of their God,/God the unforgetting, the dear God that loves the peoples/For whom He died naked, suffering shame."
13. See Des Wilson: *An End to Silence*, Cork, Carbery Books, 1987.
14. See Eamonn McCann: *War and an Irish Town*, London, Pluto Press, 1984, pp34-5.
15. For the analysis of extra-institutional protest see D E Apter and Nagayo Sawa: *Against the State*, Cambridge, Harvard University Press, 1984.
16. Or more to the point, they believe that the Republican leaders simply wish that the Northern Irish question would disappear.
17. See, for example, *Questions of History*, written by Irish Republican prisoners of war, and published by the Sein Fein Education Department, 1987. The introduction says: "It is a valuable historic document as well as a practical educational tool. The prisoners are using our history as a vehicle for influencing our present and future. In doing so they are letting us see the ideas and values that motivate them as well as the

spirit that says everyone has a part to play in the Irish Revolution, whether with a rifle or a pen."

18. Bobby Sands was elected to the British Parliament while serving a term in Long Kesh. He went on hunger strike on March 1, 1981 and died sixty-six days later on May 5, 1981. He wrote most of his poems on toilet paper, which was then smuggled out of the prison. The tradition of the hunger strike began just after World War I when the then Lord Mayor of Cork, Terence MacSwiney, was picked up for possession of seditious documents. He refused to take food for seventy-four days and died a martyr, with the Archbishop of Southwark celebrating Mass publicly for the repose of his soul. See Ulick O'Connor, Introduction to Bobby Sands: *Skylark Sing Your Lonely Song*, Dublin and Cork, The Mercier Press, 1982. See also David Beresford: *Ten Dead Men*, London, Collins Publishing Co, 1987.

19. See Mary Douglas: *Natural Symbols*, New York, Vintage Books, 1970, pp28-30.

20. The notion of legal and illegal parties is a slippery one. Recognised representatives of Sinn Fein have been elected to the British Parliament, like Bernadette Devlin or Bobby Sands. Stormont, the parliament for Northern Ireland, was abolished and replaced by direct rule.

21. They know very well, for example, that their extremism, what they consider to be their will, has lost them much sympathy and support in the Irish Republic. Indeed, their prospects for joining the South are less than at any previous time. Not a few feel deserted, repudiated, rejected by those they regarded as their fellow countrymen. The 1985 Agreement between the Republic and the British government, while a continual source of outrage for both sides (the Protestants as well as the Catholics), is regarded as a sell-out and a ganging-up, and the ultimate violation by the Republic of its IRA origins and obligations. For many it was the ancestral struggle of Southern Ireland that was the main inspiration for the IRA today.

22. Wolfe Tone, their patriarchical figure, was a Protestant, and he fought the first battles not in Ireland but in Canada. Within the retrieved myth of Irish nationalism one finds all the ingredients of a primordial nationalism, violation by outsiders, a fall from grace at the hands of alien occupiers, or, to put it in somewhat different terms, a ritualised record of displacement and dispossession, ghettoisation and diasporisation.

23. Gerry Adams: *The Politics of Irish Freedom*, Wolfeboro, New Hampshire, Brandon, 1987.

10 The Anglo-Irish Agreement: Placebo or paradigm?

PADRAIG O'MALLEY

The road to the Agreement

In the 1983 British general election Sinn Fein, the electoral arm of the Irish Republican Army (IRA) and as such the extremist public voice of unconstitutional nationalism, won 43 per cent of the Nationalist vote in Northern Ireland. It captured 15 per cent of the overall vote, compared to 18 per cent gained by the moderate Nationalists, the Social Democratic and Labour Party (SDLP).

Thus when Irish Prime Minister Garret Fitzerald met with British Prime Minister Margaret Thatcher at Chequers in November 1983, he made the case for some dramatic political initiative to alleviate Catholic alienation in the North. If it was not forthcoming, he warned, Sinn Fein would become the majority voice of the minority community. If that were to happen, it would signal the end of constitutional politics in the North, having consequences that would spill over into the South and possibly destabilise constitutional politics there. That in turn would have serious repercussions for Britain.

Despite the lack of success of attempted political initiatives throughout the 1970s and early 1980s, the political formula for an agreement was already in place. Successive Irish governments accepted that the status of Northern Ireland would not change without the consent of a majority of the people there, while successive British governments acknowledged that an Irish dimension existed, and that a devolved government would have to have the support of the Nationalist community.

In two crucial respects, however, the capacities of both governments — but especially the British government — to translate good intentions into political actions were severely circumscribed by the entrenched, unmovable positions of their respective clients.

The Unionists, secure in their constitutional position under the Northern Ireland Constitution Act of 1973 and tenacious in their belief that their numbers alone precluded them from being coerced into any form of devolved government that did not countenance majority rule, or any North-South relationship that involved more than mere "neighbourliness", were in a position to veto every proposal. Moreover, since their position on an Irish dimension was absolute, any coupling of devolution that would require their making concessions to the South made progress on devolution impossible.

On the Nationalist side, the refusal of the SDLP to enter into any discussion of devolution without a prior undertaking that an Irish dimension was an issue of at least equal standing, gave it, too, a veto power that led to paralysis.

Accordingly, the British government's power to move the political parties in the North in the direction of an accommodation was severely curtailed. It was a zero-sum game. Anything that appeared to be acceptable to Unionists was a sufficient reason for its rejection by Nationalists, and conversely, anything that appeared to be acceptable to Nationalists was a sufficient reason for its rejection by Unionists.

However, the two governments had gained some latitude by means of the Anglo-Irish process initiated in May 1980 by Irish Prime Minister Charles Haughey and Margaret Thatcher. A series of summit meetings followed and, in November 1981, both governments agreed to establish an Anglo-Irish Intergovernmental Council to give institutional expression to "the unique character of the relationship between the two countries".[1]

The summit held on 15 November 1985, at which Thatcher and FitzGerald signed the Anglo-Irish Agreement, was, according to the communique which followed it, "the third meeting of the Anglo-Irish Intergovernmental Council to be held at the level of Heads of State".[2]

The Agreement

The Agreement, ratified by Dail Eireann by 88 votes to 75, and by the House of Commons by 473 votes to 47, and registered under Article 102 of the Charter of the United Nations on 20 December 1985, effectively gave Dublin a consultative role in how Northern Ireland is governed.

It is succinct, its brevity almost concealing the craftsmanship that went into its wording. First, both governments affirmed that any change in the status of Northern Ireland would come about only with the consent of a majority of the people of Northern Ireland. Both governments recognised that at present the Unionist majority wished for no change in its status. Both governments promised to introduce and support legislation to secure a united Ireland if in the future a majority of the people in Northern Ireland were clearly to wish for and formally consent to it.

Second, the two governments agreed to set up an Intergovernmental Conference, whose functions would pertain to both Northern Ireland and the Republic of Ireland and would apply to political matters, security arrangements, the administration of justice, and the promotion of cross-border co-operation. A provision specifying that "determined efforts shall be made through the Conference to resolve any differences" — a binding legal obligation with precedent in international law — ensures that the Irish government's role is more than merely consultative even though less than fully executive.

Third, both London and Dublin support the idea of a devolved government to deal with a range of matters within Northern Ireland that would command "widespread acceptance throughout the community". Should this occur, Dublin would, nevertheless, retain a say in certain areas affecting the interests of the Nationalist minority (such as security arrangements and human rights). If devolution did not come to pass, then Dublin would continue to have a say in all matters that affect Nationalists. Finally, the workings of the Conference would be reviewed after three years "to see if any changes in the scope and nature of its activities are desirable".

Thus the logic of the Agreement and the ordering of the priorities were as follows: first, work out the relationship between the two governments on a government-to-government basis; develop a set of institutional arrangements not susceptible to the shifting vagaries of political actions in the North; and then look for an internal settlement within Northern Ireland. And thus, since widespread Unionist opposition to the Agreement was anticipated, the inducement the Agreement provides to encourage Unionists to negotiate an acceptable form of devolution with Nationalists. On the one hand, there is the carrot: the more willing Unionists are to share power with Nationalists, the smaller the role of the Conference, and hence the smaller the role of the South in the affairs of the North. On the other hand, there is the stick: the longer Unionists refuse to share power, the larger and more long-lasting the role of the South in the affairs of the North. In this sense, the Agreement was designed to undermine Unionist intransigence.

In Northern Ireland, Nationalists overwhelmingly supported the Agreement and Unionists overwhelmingly rejected it. With the passage of time, however, Nationalist support has eroded, since the Agreement has made little difference in the day-to-day lives of Catholics and has failed to deliver on some of the more conspicuous promises of reform (especially in the area of the administration of justice) that were made at the time of its signing.

Meanwhile, Unionist opposition has remained firm. One poll found that only 16 per cent of Catholics believed that the Agreement had benefited the minority community, while an overwhelming 81 per cent of Catholic respondents could find no benefit to their community from it. Protestants, of course, found even less in the Agreement with which they could identify: 85 per cent of Protestant respondents believed that Protestants had not benefited from the Agreement and only four per cent could point to some benefit to their community.[3]

The Agreement and Catholic Ireland

The Irish government and SDLP leader John Hume went to extra-ordinary lengths to show that the Agreement vindicated the wisdom of constitutional nationalism (which the SDLP represents), that the path of non-violence yielded more dividends than the path of violence, that constitutional nationalism, rather than being ideologically bankrupt, had proven itself superior in both tactics and strategy to militant Republicanism.

This obsession with securing the SDLP's base of support, with isolating Sinn Fein, overrode every other Irish consideration. At the heart of it was the explicit belief that the IRA posed a far greater threat to the Irish state than to the British government, and the implicit belief that support for militant Republicanism was the product of circumstance rather than conviction and ideology, that the IRA exploited the Catholic community rather than reflecting its ingrained essence. Thus the community could be detached from the IRA; the community could be "educated" to see its violence as wrong and perpetuating cycles of violence. The community could be wooed with a package of incentives that would be sufficient to convince it that continued support of the IRA was against its self-interest. The community could be turned against the men of violence who were exploiting it for their "evil" purposes.

Since the violence of the IRA was by definition "bad" violence, this fed the belief that the Catholic community which supported it did so mistakenly, or because they were somehow forced to do so, or because the perceived injustices of British rule make it the most palatable alternative open to them. In short, support had nothing to do with conviction.

The Agreement, the signatories believed, would detach Catholics in Northern Ireland from their current infatuation with Republicanism once they saw that the South vigorously pursued its role as the guarantor of Catholic interests, that their Irish identity had an equal standing with the Protestant identity, that their aspirations to a united Ireland were now legitimate; once the Flags and Emblems Act — which prohibited displays of the Tri-colour of the Irish Republic — was repealed; once the police were more closely supervised and their practices made less repressive in the Catholic community; once the hated UDR (a British army unit recruited exclusively in Northern Ireland and which is also almost exclusively Protestant) was brought to heel and the Diplock Courts (one-judge, no-jury courts which were used to try persons charged with political/terrorist offences) were overhauled; once the Intergovernmental Conference — and especially the Secretariat in Belfast — conveyed to Catholics that they were somehow "in union" with their co-religionists in the South.

In short, you could enumerate Catholic grievances, develop a specific strategy to alleviate each of them, and at some point the overall level of grievance would diminish to the point where the community would turn its back on the IRA. You could undermine the ethos of militant Republicanism not by defeating it but by treating its symptoms.

But history says otherwise. Alleviating the grievances of oppressed groups rarely results in their being satisfied; concessions become irrelevant, once grievances are addressed, and serve only as a springboard for further demands. "Improved conditions," writes the political psychologist Jeanne Knutson, "not only permit anger to arise; they equally create an emotional state of severe anxiety. As emotions are rekindled, the victim relives the terror of his victimisation at full intensity, and realises, for the first time, what might have happened. One never erases the identity of victim: those who live through the first injustice forever remain wary of the next attack by the identified source of aggression. Even if the aggressor loses his political, social or personal powers over the victim's life, fears are diminished, but never eradicated totally: a life-preserving, primitive belief in personal safety has been breached. Once having been so terrorised, a victim thus simultaneously grieves over the past and fears the future." Moreover, with the emergence of hope, "victims put aside ingrained feelings of traditional system support and the burdens of learned helplessness".[4]

Ironically, Catholic Ireland's attitude toward Northern Protestants was in many regards a mirror image of its attitude to Northern Catholics. It held that all things pertaining to Unionism — its Britishness, Protestantism, desire to maintain the higher standard of living that the economic link with the United Kingdom provided — could be compartmentalised and examined microscopically. Unionism had no organic essence; the sum of its parts was less than the whole. Opposition to a united Ireland was based on the Protestant desire to maintain a position of privilege rather than any deep-rooted ideological conviction.

There is a broader process at work. Catholic public moral values are not the result of convictions arrived at through a process of consensus or the secular ethic of individual responsibility. They are a product, to a considerable extent, of the Catholic Church's imprint on the entire socialisation process.

They are immutable, fixed; they make no concessions to current orthodoxies. They are handed down from one generation to the next, their eternal truths promulgated by a hierarchy that is a compulsive guardian of Catholic values in the face of every threat, in constant fear of losing its power and position, unwilling to entertain any idea that might erode its authority. They are the teachings of the One, True, Holy, Catholic and Apostolic Church. They are written in stone, they are enduring, permanent, beyond question. Because the Irish Catholic moral value system is based on edict rather than arrived at through the consensual development of shared values, Catholics are not easily able to respond to value systems that evolve from a community's participation in the process that leads to the propagation of such shared values. Because their own value system often lacks the integrity of their convictions, they tend to dismiss the integrity of others' convictions.

Underlying Catholic Ireland's interpretation of the New Ireland Forum Report and the Anglo-Irish Agreement, is its attitudes toward Protestantism. Catholicism is right, its values God-given; a united Ireland is historically mandated, a matter of inevitability. Protestants belong to the "wrong" religion; Unionists to the "wrong" state. Protestants should be converted to Catholicism, Unionists to a united Ireland.

Acknowledgement of Protestant beliefs does not imply acceptance, approval, or concession of their legitimacy; acknowledgement of Unionism does not imply acceptance, approval, or concession of its legitimacy. The willingness to allow Protestants their Protestantism does not diminish the obligation you have to get them to see the errors of their ways; the willingnss to allow Unionists their Britishness does not diminish the obligation you have to convince them that they are really Irish, or, at the very latest, that their long-term interests would be more adequately safeguarded under some all-Ireland umbrella than in a British state.

Hence the logic of the Agreement that prevailed in Catholic Ireland and, one should add, to a lesser but nonetheless significant extent in British government circles. A diminishing level of IRA violence was coupled with diminishing Protestant opposition to the Agreement. Once it became clear to the Protestants that the constitutional status of Northern Ireland would be unaffected by the Agreement, that their real interest in remaining part of the United Kingdom was adequately protected, that the Intergovernmental Conference was a reality both governments would not back away from, it was assumed that their opposition would decrease.

Protestants would acquiesce in the Agreement, even if they would not endorse it. Self-interest would move them to share power with Catholics, in order to minimise the purview of the Conference. Power-sharing would further reduce support for the IRA since it would give Catholics an active voice in government itself. Protestant opposition would further diminish with increasing Catholic support of and membership in the ranks of the police. And, as their opposition eroded and their distrust dissipated, they would become increasingly responsive to new and more extended forms of co-operation between North and South.

The Agreement is a superb exercise in the power of rationality: state the objectives and define the measures to implement them. It argues that specific grievances are more significant than the essence of grievance itself, and reflects the belief that tinkering with governance arrangements can override the differences embedded in the mutually exclusive ideologies of both communities, or at least accommodate them sufficiently to undermine militant Republicanism (the aim of the Irish government), or to defeat terrorism (the aim of the British government).

However, since the Agreement had to be simultaneously capable of accommodating Catholics without permanently alienating Protestants, it had to reflect this duality of intent. Thus its ambiguity; its emphasis on intent rather than fact; its adherence to process rather than specifics; its design as a framework rather than as a set of provisions; its malleability rather than its rigidity; its indirectness rather than its directness.

The Agreement and Protestant Ulster

In the ambience of interpretation and counter-interpretation, Protestants took the only course they knew. The translated the Agreement into the language they understood; they cut through the ambiguity, the nuances, the significant omissions, the balance and order of the arrangements it proposed, and got to the heart of the matter. The Agreement mandated Dublin Rule.

Protestant reaction was one of undiluted rage, rage that came from the wellsprings of their beings, an expression of anger and virulence at the British government so profoundly felt that it altered for all time the relationship between Protestant Ulster and Britain. The language of resentment used a vocabulary of hate. Mrs Thatcher was "a Jezebel", "an unprincipled, shameless hussy" who had signed away Ulster Protestants' "inalienable right to citizenship in the UK".

"We pray this night," Ian Paisley intoned to his congregation at Martyr's Memorial Church on the Sunday following the signature of the Agreement, "that Thou wouldst deal with the Prime Minister of our country. We remember that the Apostle Paul handed over the enemies of truth to the Devil that they might learn not to blaspheme. In the name of Thy blessed self, Father, Son and Holy Ghost, we hand this woman, Margaret Thatcher, over to the Devil that she might learn not to blaspheme . . . Oh God, in wrath take vengeance upon this wicked treacherous lying woman, take vengeance upon her, O Lord, and grant that we shall see a demonstration of Thy Power."[5]

In the Protestant perspective, the one element of the Agreement that counted was the Intergovernmental Conference. The Conference was viewed not as a small co-operative gesture but as a coalition of a government in embryo. In the Unionist view, the language in the Agreement requiring "determined efforts to resolve differences" meant that Dublin would get its way 50 per cent of the time. The Conference was seen as the first step toward an all-Ireland state.

But Protestants reserved their most vituperative reactions for something that was not a part of the Agreement — the Secretariat's location in Belfast. This represented the greatest threat of all. The spectacle of Irish ministers coming to Belfast, the reality of Irish civil servants being permanently located there, struck the raw nerves of their most profound fears. The Secretariat was the bridgehead, it penetrated the fortress, allowed the enemy to secure a foothold. "Dublin," said the Democratic Unionist Party's Ivan Foster, "not only [had] its foot in the door — the door was off the hinges."[6]

Once again Ulster Protestants found themselves in near total isolation. The international community received the Agreement with acclaim, and the British political parties — in a rare display of unanimity, and being more than willing to have Britain's Ireland problem become Ireland's Ireland problem in some more obvious and visible way — gave it their unconditional imprimatur. The Unionists' plangent cries that they had not been consulted reverberated in the dead air of the political wasteland they inhabited, their apoplectic denunciations of the Agreement, Mrs Thatcher, the South — anything that came within range of their free-floating anger — obfuscating their cause for just complaint.

They saw themselves as the real victims of British misrule, deprived citizens in their own state. They were the ones whose birthright had been denied when their government was taken from them in 1972; they were the ones who had been brought to heel by Direct Rule; they were the ones who were being informed by dictatorial decree that the Act of Union of 1801 — which, in their view, had conferred on Protestants an unalienable right to be British, ratifying a covenant made with the Crown, the symbol of continuity and perpetuity of order, rather than with the government of the day — had been abrogated. They were now conditional members of the UK, only part of it for as long as that was the wish of a majority — a far cry from Mrs Thatcher's assurance in 1979 that Northern Ireland was as British as Finchley (the constituency in North London that Mrs Thatcher represents in Parliament). The goal of a united Ireland was now within the grasp of the IRA. Violence pays.

The Agreement impugned the one historical exigency of Protestantism — to uphold the settlement of 1689, to defend Protestant civil and religious liberty. It violated their mythic conception of self, their idealised experience of how they had emerged as a community, of how the Glorious Revolution had led to liberation from spiritual tyranny, the birth of the individual, the triumph of light over darkness, and, as James McEvoy writes, "the emergence of an ordered world of stable and finite realities — nation, crown, Parliament and property rights; in short, the final establishment of justice, under the active intervention of a favourable deity".

Since "justice and the good ordering of society, attunement to the wholeness of reality" brought with them peace and progress, industry and prosperity, "the truly ordering forces . . . could not operate for those who remained outside the framework of such attunement". Those people must remain in darkness, "in servitude to other forces, and suffer the consequence of ill-attunement to reality: poverty and backwardness". Thus, basic to the Protestant interpretation of self was "the belief that the Protestant version of Christianity was inherently superior, both in itself and in the attitudes and actions to which it gave rise, to the Catholic Church," that they were to rule and not be ruled.[7]

In their anger and terror, they invited substantiation of the truths of their own myths, desperately searching for the ritualistic acts that would recapture the primordial experience of order that had initiated their history. They were seeking the Protestant equivalent of the Republicans' hunger strikes, a form of protest so fused with the elements of the past, so evocative of their inner realities, that it would free them from the strictures of their predicament and provide them with a way out.

Action had its parallels in the past: mass demonstrations in Belfast and the burning of Mrs Thatcher in effigy to evoke the mass demonstrations of 1912 against Home Rule and the burning in effigy of Herbert Asquith; the Days of Action, bringing the province's industry and commerce to a standstill, to evoke the Ulster Workers' Council strike in 1974, which had brought down the power-sharing executive and the Sunningdale Agreement; the 500 000 signatures on a petition to the Queen calling for a suspension of the Agreement to evoke the Ulster Covenant in 1912.

"Ulster Says No" became the catchcry to rally the faithful. "Iron Lady be warned," their banners cried, "your iron will melt from the heat of Ulster." They would never accept the Agreement. They would withdraw from Westminster, from local councils, from regional bodies. They would make Northern Ireland ungovernable. They would show that Northern Ireland could not be governed without the consent of the majority, and if their withdrawal from government meant that they ceded ground to the Protestant paramilitaries, the fault would not be theirs; responsibility for the sectarian violence that might result would lie with the two governments.

The two main Unionist parties — the Official Unionist Party and the Democratic Unionist Party — closed ranks, the divisions between them, their competition for the hearts and minds of Ulster's Protestants, put aside. Once again "the myth of Unionist siege [would have] its greatest social utility as an expression of a primary solidarity of purpose within a community principally defined by a determination to resist".[8] Their sheer determination not to accept the Agreement under any circumstances or to engage in talks with Nationalists unless the Agreement was suspended would, they believed, ultimately prevail. They resigned their seats in the Westminster Parliament, calling the special elections held to fill them a referendum on the Agreement.

However, they lost one seat to the SDLP, and the total "No" vote fell under the 500 000 mandate they had sought. But they persisted. They had a simple objective, no grand design. In their fury they did not care whether their actions weakened their union with Great Britain. They were driven by their age-old fears: any association with the South would lead to the ineluctable absorption of the Protestant people of Ulster into an all-Ireland Catholic state. But the world had changed. The rituals that had once imparted an inner reality were now merely chimeric. They had lost their force.

An assessment

Ultimately, of course, the Hillsborough Agreement will be judged on the extent to which it achieves its avowed aims — that is, the extent to which it promotes peace and stability in Northern Ireland and helps to reconcile the Protestant and Catholic communities with their divergent but legitimate interests and traditions. The notion that these aims can be achieved, however, was the product of explicit and implicit assumptions on the part of both Dublin and London — assumptions that are, perhaps, not entirely tenable.

Unfortunately, even if the new political arrangements successfully addressed Catholic concerns and support for Sinn Fein began to diminish, this would provide no guarantee of peace, stability, or a reduced level of alienation between Catholics and Protestants. Indeed, it would appear that some reforms may have had an effect opposite to the one intended. Reforms attributed to the Agreement by the SDLP may have weaned Nationalist votes away from Sinn Fein, but there is no reason to believe that this resulted in a decrease in the activities of the IRA. There is no *necessary* relationship between the capacity or will of the IRA to commit acts of violence and the level of political support for Sinn Fein.

On the contrary, the IRA has been able to step up its campaign of violence. In each year since the Agreement went into effect the level of IRA violence has exceeded its pre-Agreement levels. Each new killing of a member of the UDR or RUC has only strengthened the conviction of Protestants — who already see themselves as the victims of a calculatedly cold-blooded campaign of genocide conducted by the IRA — that the Agreement has facilitated and encouraged the IRA death machine. In fact, the Agreement has not been a cure-all for Nationalist grievances and its tangible benefits are few.

The Agreement has had no significant impact on the level of violence.

The number of killings in 1986, 1987 and 1988 were above 1985 levels. The IRA still strikes randomly, ruthlessly and with little regard for life; the Loyalist paramilitaries are again engaging in the random assassination of Catholics and the divisions between the two communities remain as great as ever. So, relying upon facts and figures and the more obvious measurements of progress — and always bearing in mind that what is progress to one community is more often than not an anathema to the other — the Agreement has hardly lived up to its early billing. Indeed, in terms of the extent to which the Agreement has achieved its principle aims, the Agreement must be considered a failure.

However, to judge it in these terms alone would be to misunderstand its real purpose and the profound impact it has had on the Unionist psyche.

There can be no doubting the Agreement's historic significance. For the first time since 1920 when the partition of Ireland occurred, the British government explicitly recognised that the Republic of Ireland has a role to play in the governance of Northern Ireland — a far cry from the 1982 declaration by the Foreign Office that Britain considered itself "under no obligation to consult the Dublin government about matters affecting Northern Ireland".[9] Giving Dublin a role in Northern Ireland constitutes an implicit acknowledgement by Britain that the partition of Ireland, in political and social terms, has been a failure.

For its part, the Irish government was prepared to explicitly accept the fact that Northern Ireland will remain within the United Kingdom as long as that is the wish of a majority of the people there. This amounts to an implicit acknowledgement that unification is an aspiration rather than an inevitability. The Agreement, therefore, is a quid pro quo of sorts. In exchange for the Irish government's recognition that Unionists have the right to say "no" to a united Ireland, the British government was prepared to give the Irish government a role in Northern Ireland in areas relating to the aspirations, interests and identity of the Nationalist minority. It was prepared to accept, in other words, an Irish dimension.

Accordingly, the status quo that had existed from 1920 to 1985 was destroyed with two strokes of the pen, and if the Union as Unionists knew it was over, so too were the fanciful notions — so earnestly and witlessly promulgated in the South for sixty years — that only Britain's presence in Northern Ireland stood in the way of Irish unity. In this sense, the old order is dead, the Agreement being, in some respects, the plug that was pulled on the artificial, political life-support systems that were sustaining Northern Ireland, and in other respects, a shock to the body politic itself.

Despite the fact that the Agreement explicitly states that there will be no change in the status of Northern Ireland as part of the United Kingdom without the consent of a majority of its people, Unionists believe that the Union — that is, the connection between Northern Ireland and Great Britain dating from 1920 and beyond that to 1801 — is over. They further believe that the Agreement gives the Irish government a toehold in the North that is, in their view, the first step on the "slippery slope" to unification.

These perceptions have undermined one component of their identity — their Britishness — and strengthened another — their Protestantism. This, in turn, has aggravated the enduring tension within Unionism itself: between those whose primary interest lies in maintaining the connection with Britain and those whose primary concern lies in not becoming part of an all-Ireland state. The Agreement has undermined the basis of that loyalty.

Before Unionists can engage in substantive talks with Nationalists about future political arrangements for the North, they must decide what their relationship to Britain means to them and what the continuing form and extent of that relationship is to be — whether they want to patch up the marriage, work out a new relatonship, or settle for a divorce. It may be a long, painful, and ultimately even an unsatisfying process, but until they resolve these questions, the issue of talks with Catholics is moot. The Agreement, therefore, is a catalyst which, by requiring Unionists to redefine their relationship to Britain, will, by necessity, require them to redefine their relationship with the rest of Ireland.

As far as the constitutional Nationalists are concerned, the Agreement assumed that the SDLP wanted power-sharing and believed it could work. However, the SDLP has since made it clear that it has "no ideological commitment to devolution".[10] Moreover, even though the Agreement has been in operation for four years, there is still no indication that the SDLP has any real idea of what power-sharing arrangements it might find acceptable in regard to the powers and functions it might be required to share, and the trade-offs it would be called on to make in the form of a lesser involvement on the part of the Republic: a diminution in the authority of the Conference, and the possible removal of the Secretariat, with its complement of Irish civil servants, from Belfast.

The SDLP, therefore, has to settle on an agenda that spells out what it wants and what it will settle for in terms of participation in the government of Northern Ireland vis a vis its aspiration to Irish unity. It has to develop a policy that balances its requirements for power-sharing with its commitment to Irish unity and the expression of that commitment in the form of the Irish dimension.

Sinn Fein has not quite figured out where it is going, even if its electoral tactics are successful; it has no idea where it is going if its tactics are unsuccessful. Success at the ballot box justifies Sinn Fein strategy; failure ultimately will bring into question the legitimacy not merely of the movement itself but of the armed struggle it advocates. Participation in constitutional politics, once undertaken, cannot be abandoned, since the cost of doing so would appear to be too great. Failure at the ballot box, however, especially in the South, may involve a cost that is just as great.

Sinn Fein has struggled to find a new voice in the 1980s, a proper balance between its new-found pragmatism and its old-time righteousness. It has struggled to reconcile its contradictory impulses — the adoption of the language of the democratic process and the sanction of the ballot box, and the simultaneous assertion that no argument, no countervailing constitutionalism, could derogate from the IRA's absolute right to conduct the armed struggle. The people were right — but only up to a point. Their wishes were paramount — but only when they had the right wishes. They did not have the right to be wrong on the national question. Sinn Fein wanted "a sovereign, independent, united Ireland".

At the core of the new strategy — "armed struggle in the six counties in pursuance of British withdrawal, and political struggle throughout the whole thirty-two counties in pursuance of the Republic," is how Gerry Adams articulates it[11] — was a contradiction. The greater the commitment to armed struggle in the North, the less the appeal of Sinn Fein to voters in the South who were being asked to sanction "bad" violence. The poorer the performance of Sinn Fein in the South, the more it would expose the IRA's lack of legitimacy. The drive for success at the ballot box would create pressures to subordinate the armed struggle to political pragmatism; conversely, lack of success would simply reflect the real absence of support for the IRA's campaign of violence. "Without politics, you may be able to bomb and shoot a British connection out of existence, but you will not bring anything into existence," writes Gerry Adams.[12]

But what if the violence makes the politics impossible? What then? In the final analysis, Sinn Fein's identity is inseparable from its relationship with the IRA. Without that relationship, it is little more than a left-leaning, essentially working-class party with a limited constituency and of limited consequence. With that relationship, it exerts an influence on events out of all proportion to popular support for its policies. The power of the movement comes out of the barrel of a gun.

Final thoughts

Ultimately, however, the success of the Anglo-Irish Agreement will be measured in terms of the extent to which it facilitates a resolution of the often conflicting agendas it poses for the major constitutional players. It may fail to do so. The extent of the segregation of the two communities is almost total; the Agreement at best treats the symptoms and not the causes of that segregation; and the best-intentioned structures developed and imposed from without cannot substantially alleviate the wellsprings of a seemingly intractable and immutable communal division that emanates from deep within.

The conflict is about permanent identity-in-opposition, requiring every occasion to splice itself — like some feat of genetic engineering — into its tribal components, precluding a shared sense of grieving, the rhetoric of exclusion accommodating only a mutual sense of betrayal. It is the conflict itself which provides the meaning, the structures of understanding and purpose that prevent social disintegration. "[The] ideology of conflict," the sociologist Peter Marris observes, "does not press claims or demand rights only for their own sake, but to sustain the conflict itself. The intransigence wards off the unbearable strain of incorporating the contradictions, and cannot help being identified with both sides."[13]

But even if the Agreement does fail, it has altered the context for all future policy action. Its enduring importance, therefore, is the fact that it was implemented in the first place.

Postscript, 22 June 1990

In the year elapsed since this paper was written, Peter Brooke, who became Secretary of State for Northern Ireland in July 1989, conducted a series of one-on-one "talks about talks" with the leaders of Northern Ireland's constitutional political parties. There is, it is now clear, some common ground.

All the constitutional parties to the conflict believed negotiations should take place. But how pre-conditions — especially how the Unionist's insistence on a suspension of the Agreement and the closing of the Conference's Secretariat in Belfast — might be finessed, who

At the core of the new strategy — "armed struggle in the six counties should attend the negotiations, the areas of discussion they should involve, how the process itself should unfold, and under whose auspices talks should be held, were questions over which there were widespread differences.

During the first six months of 1990, Brooks pursued a two-tier strategy: on one level, inter-party talks within Northern Ireland to develop new governance structures for Northern Ireland; and, on a second level, bilateral talks between the two governments to explore possibilities for a new accord to replace the Anglo-Irish Agreement, should an internal settlement among the North's political parties be reached. Brooke was cautiously optimistic, and, if all went well, inter-party talks will have occurred in the autumn of 1990.

ENDNOTES

1. Communique issued following meeting between Garret FitzGerald and Margaret Thatcher, 6 November 1981. See *Irish Times,* 7 November 1981.
2. Paul Arthur: "The Anglo-Irish Agreement: Events of 1985/86", *Irish Political Studies* 2, 1987, pp99-107.
3. *Fortnight* and Ulster Television Survey, *Irish Times,* 25 March 1988.
4. Jeanne N Knutson: *Victimisation and Practical Violence: The Sceptre of Our Times,* uncompleted book MS, 1981, pp73-6.
5. Padraig O'Malley: *Biting at the Grave: The Irish Hunger Strikes and the Politics of Despair,* Boston, Beacon Press, 1990.
6. Interview with author, 25 August 1986.
7. James McEvoy: "Catholic Hopes and Protestant Fears", *The Crane Bag* 7, No 2, 1983, pp101-2.
8. Terence Brown: "The Whole Protestant Community: The Making of a Historical Myth", A Field Day Pamphlet, No 7, 1985, p10.
9. From Mr Haughey's address to Dail Eireann during debate on Anglo-Irish Agreement, 19 November 1985. See *Irish Times,* 20 November 1985.
10. Austin Currie, former SDLP senior spokesperson, quoted in *Fortnight,* No 256, November 1987, p7.
11. Gerry Adams quoted in O'Malley: *Biting at the Grave,* op. cit., p242.
12. Gerry Adams, *The Politics of Irish Freedom,* Dingle, County Kerry, Brandon, 1986, p64.
13. Peter Marris: *Loss and Change,* London, Routledge and Kegan Paul, 1986, p96.

11 Ruling group cohesion

JANNIE GAGIANO

This chapter reports on the attitudes of white South African students as a microcosm of the white ruling group. It is based on a survey, conducted in mid-1989, of a large sample of white students.[1] The political attitudes of white student populations in South Africa yield a vantage point from which one can gauge the attitudinal trends that are prevalent and emerging within the ranks of the politically dominant white middle-class formations in the wider society.

As is the case with dominant groups in many other divided societies, the white student community in South Africa is strongly influenced by the major institutions and organisations that operate as agencies of political socialisation and mobilisation in the wider society. Student attitudes reflect the dimensions of the public ethos which underpins group consciousness and coherence. If the mounting pressures directed at the white ruling group in South Africa are undermining its political will and determination, it is among white students that one is as likely to observe it as anywhere else in society.

The whites are constituted as a political community because of their differential and superior access to the levers of power in the state. They also share a vague notion that because of a common European ancestry they must stand together to protect Western values. Otherwise they reflect fragmented political culture. Although the contours of the deep political divisions that historically separated the English- and Afrikaans-speaking groups have softened somewhat, they remain. (They are clearly observable in the ranks of the younger generation, like the students in our sample.) Within the white group the Afrikaans-speaking community holds a dominant position in the political arena and its members occupy most of the decisive command posts in the administrative systems of state. The Afrikaner community can be presented as the vital core of the overall structure of white political supremacy.

In the survey findings there is a striking difference between the attitudes of white English-speaking students and Afrikaners. The reason for this is that the context in which white South Africans form their political attitudes and the communal environment in which political socialisation and intercourse takes place is decisively shaped not only by the effects of racial segregation but also by language group attachment. Language group education, the maintenance of institutional pluralism across the language groups, and the decisive influence these institutions have in shaping political loyalties and perceptions (especially among the Afrikaners) tend to reproduce a fragmented political culture and provide a context in which political socialisation occurs.

For example, the typical Afrikaner will go to an Afrikaans school, live in a suburb where other Afrikaners are well represented, attend the service of one of the family of Afrikaans churches, rely on an Afrikaans language newspaper for political information, go to one of the six Afrikaans universities (if he/she is clever or the parents are rich), and work for the state or for a firm with an Afrikaans ambience, like the insurance giant Sanlam or the Rupert group of companies. This pattern would be replicated, in reversed form, for English-speakers.

We offer some brief evidence from our survey to make the point. More than 90% of Afrikaners in our sample belong to one of the "family" of Afrikaans churches (74% are members of the Dutch Reformed Church) while less than 1% of English-speakers belong to these churches. Equally, less than 2% of Afrikaners belong to any of the denominations which predominate among English-speakers. Ony 1,9% of Afrikaans-speaking students in the sample study at English language medium campuses and only 12,7% of English-speakers study at Afrikaans language medium universities. More than 80% of Afrikaner students belong to the two parties on the right and centre right of the political spectrum, the Conservative Party (CP) and the National Party (NP), whereas more than 80% of English-speaking students belong to the Democratic Party or political formations to the left of the Democratic Party.

It is one of the conventional wisdoms of revolutionary theory that if a ruling group can maintain its internal solidarity and cohesion, other things being equal, it has a greater chance of successfully resisting displacement by revolutionary forces. An influential voice in con-temporary theorising about revolutionary change, Theda Skocpol, points to the importance of the attitudes of the regime's own cadres. "In any event," she argues, "what matters most, is always the support or acquiescence not of the popular majority of society but of the politically powerful and mobilised groups, invariably including the regime's own cadres".[2]

Referring to the specific circumstances in South Africa, Samuel Huntington has argued that revolutionary violence does not have to be successful to be effective. "It simply has to create sufficient trouble to cause divisions among the dominant group over the ways to deal with it. At this point, one of the three stanchions (white political unity) of the existing system begin to disappear, and the ability of the government ruthlessly to apply its instruments of coercion becomes compromised."[3]

These views raise some of the questions which the evidence gleaned through our student survey can help to answer. Can one observe the emergence of feelings of guilt about the continuation of a system of government that preserves privileged places for whites in its command structures? To what extent is the government or its agencies of social control criticised, censured or even obstructed by members of the ruling group when it sets about repressing expressions of political dissent emerging from the popular sectors of society?

Group cohesion amongst whites, as reflected in the ranks of white students in South Africa, can be observed along a number of dimensions. One of these is the degree of cohesion internal to the white group. Another is cohesion internal to the dominant Afrikaner group. A third is white cohesion in the face of a revolutionary challenge to the South African state from the major liberation movements. It is on the latter that this chapter will concentrate.

Attitudes towards the state and its challengers

We shall take the dimensions of white group cohesion to be more or less revealed by the patterns of affective attachment to the institutions, policies and practices of the state. The breakdown of that cohesion can be taken as revealed by the extent to which an attachment to the organisation, policies and practices of the major challengers to this state is emerging and reflected in the attitudes of our respondents.

To show the patterns of attachment to the state we shall use four different indices. The first of these is one that measures affect towards institutions located within the state but which harbour opposition to the principal policies of the ruling National Party. The second one measures affect towards institutions located within the state but which harbours opposition to the principal policies of the ruling National Party. The third taps feelings towards the security establishment and the fourth indicates acceptance or rejection of the major contenders in

South Africa with claims to rule the country.

The differential degree of attachment to the state in the white community is very evident if one compares Afrikaners to English-speakers.

Table 11.1: Sympathy/antipathy towards public authority

	Afrikaans-speaking whites %	English-speaking whites %
Very sympathetic	23,8	2,7
Sympathetic	49,8	18,7
Apathetic	20,4	37,5
Unsympathetic	5,1	32,6
Very unsympathetic	0,9	8,5

To account for this difference one has to consider that Afrikaner political culture was decisively moulded by the ethnic nationalism that took possession of their collective political imagination for the greater part of the twentieth century. To capture control of the South African state and invest it with a distinct Afrikaner character was one of the major goals driving their political project. All the major organisations within the Afrikaner community were locked into this project. Through their hold on their members they fostered feelings of legitimacy towards the Afrikaner-controlled state, its symbols and its institutions.

Capturing power through the National Party in 1948, the Afrikaner community appropriated the state as their own, like one would establish ownership of a farm *(boereplaas)*. In a previous article I called this the *boereplaas* syndrome. Thus it is very common for Afrikaners to speak of "our" Minister of this or that, "our" police force, "our" defence force and so on, while it is very rare for English-speakers to do so. Much of the disaffection observable among English-speakers relates to a feeling of exclusion from the inner precincts of the state and from a lack of identification with its pro-Afrikaner symbolic character. As we shall show later, this disaffection does not convert into a significantly favourable disposition towards the African National Congress (ANC) or the United Democratic Front (UDF), for example.

However, the South African state does not only embrace those institutions visibly controlled by Afrikaners. It also includes institutional arrangements that incorporate the homeland system of government for blacks, the House of Delegates for Asians, the House of Representatives for coloureds and the Inkatha movement.

These are called the institutions of co-optation. Table 11.2 gives the attitudes of white students towards them.[4]

Table 11.2: Feelings towards the co-optation institutions

	Afrikaans %	English %
Very sympathetic	11,9	10,7
Sympathetic	43,4	38,6
Apathetic	34,9	36,0
Unsympathetic	8,2	12,0
Very unsympathetic	1,6	2,8

Another important indicator of support for the state is the attitude adopted towards the security establishment in the country, that is the South African Defence Force (SADF), the police and the security police.

Table 11.3: Feelings towards the security establishment

	Afrikaans %	English %	Total
Very sympathetic	46,6	5,5	28,4
Sympathetic	38,0	28,0	34,5
Apathetic	8,9	20,9	14,3
Unsympathetic	4,7	27,1	14,2
Very unsympathetic	1,9	18,1	8,6

A similar pattern can be observed in attitudes towards conscription and the refusal to do military service as a form of political protest. Among Afrikaner male students 86,3% say that they will never refuse to do military service as a form of political protest whereas only 55,7% of English-speaking males say so. Among the English-speaking students 2,3% admit that they have already refused military service, 11,3% indicate that they would do so as a form of protest and a further 24,2% say that they would consider doing so as a form of protest. Among Afrikaners only 2,7% say they would refuse to do military service and a further 7% say that they would consider doing so.

Attitudes towards the liberation movements were registered on two different measuring devices. The first was an index tapping affect in the same way as the indices we have been reporting on so far (we use a device known in the trade as a "feeling thermometer"). The second was based on answers to the question: "To what extent would you be able to live with a government controlled by each of the following movements or parties?" (followed by a list of the major contenders for political power in South Africa).

Table 11.4: Responses to a prospective ANC government

	Afrikaans %	English %	Total %
Would welcome it	1,1	6,6	3,4
Would accept it	6,6	21,7	12,9
Uncertain	4,3	15,0	8,8
Protest against it peacefully	12,1	8,5	10,6
Emigrate for political reasons	31,7	38,0	34,4
Would resist physically	44,1	10,2	29,9
n = 3842			

On our measure of sympathy towards the ANC the pattern recorded above was well substantiated. More than 90% of the Afrikaans students were unsympathetic to the ANC (5% sympathetic and a further 5% apathetic). For English-speakers 28% are sympathetic, 14% apathetic and 58% unsympathetic.

It is clear that the greater part of the Afrikaans-speaking community, largely convened in the centre right and right-wing parties, are hostile and resistent to the ANC (and the UDF). If the white community has a "soft underbelly" at which the liberation movements could probe, it would have to be located with elements in the English-speaking community (like white university students who are, as we indicated earlier, substantially displaced to the left of the English-speaking community at large) and in the ranks of the Democratic Party (again bearing in mind that students would be overrepresented in the left wing of the DP). These two categories of respondents have a large overlap.

The following table might illuminate the issue.

Table 11.5: English-speaking attitudes to prospective governments

	DP gov %	ANC gov %	NP gov %	UDF gov %
Welcome a	68,1	6,6	14,0	14,0
Accept a	16,6	21,4	42,8	22,8
Uncertain about a	10,2	15,0	4,3	23,7
Protest against a	3,6	8,5	31,5	9,4
Emigrate	0,6	38,0	4,2	22,5
Physically resist a	8,8	10,2	3,2	7,5
n = 1629				

All the liberation movements in South Africa offer majoritarian alternatives to the present system of government and the ANC and Mass Democratic Movement (MDM) offer a nation-building model conceived along Jacobin lines and in terms of which political identity is mediated through the rights of individualised citizenship in a unitary state. The building of political formations on the basis of ethnic or racial identities is deliberately proscribed and the ANC has indicated that under their rule attempts at ethnic mobilisation would be visited with the "libertory intolerance" of the state.

We presented our respondents with six different constitutional options for South Africa and asked them to indicate their preference. The following results were obtained.[5]

Table 11.6: Alternative political dispensations for South Africa

	Afrikaans %	English %	Total %
Partition	10,7	1,4	6,8
White control	26,2	3,2	16,6
Racial federation	29,9	30,5	30,2
Geographical federation	26,6	42,2	33,1
Unitary state	5,1	20,0	11,3
Worker state	1,6	2,7	2,0
Total	58,3	41,7	100
n = 3732 (The bilingual group was left out of the table.)			

From these distributions it is clear that almost 90% of white students in

South Africa prefer a constitutional dispensation for South Africa that in some way or other makes provision for the protection of group (read white group) rights. Just under 70% of Afrikaners prefer dispensations where racial forms of political representation (or domination) are built into the constitution. Only 5,1% endorse the unitary state model promoted by the major liberation movements. Among English-speakers this is not very much higher at 20%. Most of them also prefer group-based constitutional solutions.

The most endorsed option among white students in South Africa is a geographical federation. Even among Afrikaners and supporters of the National Party this option has gained in popularity. Next to the unitary state option this arrangement offers the most opportunities for black people to compete for power in the state. Given the known strong support for the Congress movement amongst blacks in South Africa, can one take this to mean the piecemeal acceptance of the eventual realisation of the state and societal ideals of the ANC? Such a conclusion would be misleading.

Group and geographically based federal schemes offer more opportunities for mechanisms that protect white interests to be built into a constitution and to offset some of the consequences of bringing blacks into a common polity. Hence the appeal of consociational type models of constitution-making among those who think of themselves as white liberals, identify with the Democratic Party and reject claims by the government that they are acting as a Trojan horse aiding the revolutionary transfer of power to the liberation movements.

We suggest that liberals see federation as serving a counter-revolutionary strategy. This is based on two beliefs that are commonly encountered in their political thinking. The first is that making the polity more inclusive and creating opportunities for political competition will enhance the liberal parameters of a polity rather than threaten them. Political instability and the associated opportunities for revolutionary entrepreneurship in South Africa are a function of the inaccessibility of the political centre to blacks — it is believed that accessibility will convert to stability. This expectation is supported by a second contention: that any ruling formation in South Africa must keep the essential capitalist sector intact on pain of destroying or fatally wounding the famous "goose that lays the golden eggs".

The liberal emphasis is in support of the gradual liberalisation of the polity, the desegregation of society and the revitalising of the economy by reallocating resources away from control by the state to sites where they can generate growth under capitalist management. The counter-revolutionary objective of this strategy is to deradicalise black elites (also in the ANC) by co-opting them into the capitalist economy, thereby reducing the salience of politics that seek to control the state as a means to compete for the material rewards of society.

Hence South African liberals offer a decentralised polity run on federal lines to protect middle-class political influence and economic privilege against the prospective demands of a central government responding to the pressures of a populist majority. It is pro-West and strives to lock the economy firmly into the ostensibly benign embrace of the First World economies.

Attitudes to violence, protest and repression

The position people have adopted on the role of violence in the process of political change in South Africa has been a useful rule of thumb to decide their attitude towards a possible revolutionary transfer of power in South Africa. The ANC has declared violence (principally expressed through an armed struggle) an essential element in its overall strategy to put pressure on the state. The MDM, although publicly committed to peaceful political strategies, has yet to condemn the violence generated in the black communities against some of its political opponents and some of the authorities representing the state. The MDM has tended to excuse this violence as a legitimate form of retaliation against the sometimes violent methods of repression used by state agencies.

The ANC's stand on the armed struggle has a poor reception among whites in South Africa. About 10% of the white students in our sample agree with the notion that black people in South Africa have a right to take up arms against the government. Only about 8% agree that black militant groups deserve more sympathy when they turn to violence as a means to redress their grievances. Although some 20% agree that terrorism and guerrilla action is frequently the only way in which a suppressed people can get what they deserve, more than 80% judge that the use of violence to bring about political change does more harm to the cause than good. They also have misgivings about the efficacy of violence as a political method. Only 14% agree that political violence is a way to get results and some 15% think that vital political change can be brought about through demonstrations and the disruption of the system.

On a summary index composed of the items reported above the following emerged:

Table 11.7: Justification for the political use of violence (total sample)

Violence is:	
Justified	5,8%
Uncertain	20,7%
Not justified	73,5%

The MDM has played an important role in stimulating organised acts of public protest in the black communities and on the campuses of English-speaking and black universities. These actions include protest marches, rent and consumer boycotts, mass stay-aways from schools and universities, organising industrial strike action and so on. The ANC regards the disruption of government and the mobilisation of black communities into protest action against the state, as one of the "four pillars" of their revolutionary strategy.

The South African Police, security police and the defence force have acted harshly in repressing these initiatives. People participating in protest marches have been arrested and detained, the police have regularly broken up peaceful demonstrations and rallies in the townships and on university campuses. The eviction of people who refuse to pay their rent for political reasons, the police firing at demonstrators with sharp ammunition and the summary dismissal from their jobs of people who took part in stay-aways have become commonplace.

To measure attitudes towards these developments we used three indices. The first tries to tap protest potential, that is the extent to which respondents have or are prepared to participate in acts of political protest. The second tries to measure coercive potential, in other words the extent to which respondents are prepared to justify the methods by which state agencies repress or punish acts of political protest and violence. The third probed for the extent to which violence was justified as a political weapon. The following responses were registered:

Table 11.8: Protest potential (itemised index)

	Afrikaans %		English %		Total %	
	HD	WD	HD	WD	HD	WD
Demonstrate peacefully	6	32	24	34	13	33
Join consumer boycott	3	21	10	30	6	25
Join protest march	4	12	22	12	11	12
Mass stay-aways	2	8	5	12	3	10
Refuse military service	0,2	3	1,5	15	0,7	8
Refuse to pay taxes	0,6	3	0,7	7	0,6	5
Join illegal strike	0,2	2	0,3	8	0,3	5
Damage property	0	0,2	0,6	0,3	0,3	0,3

(The percentages were rounded — HD and WD at the head of the columns indicates that respondents have done or would do the protest actions.)

A composite index of protest potential was built up from these items. (We assigned weights to the different items on the basis of the application of Guttman scaling techniques.) The results presented by language group and by party affiliation were as follows:

Table 11.9: Protest potential (composite index)

	Afrikaans %	English %	Total %
No protest potential	60,9	41,9	52,9
Low protest potential	27,0	29,9	28,2
Moderate potential	6,1	14,3	9,5
High potential	6,0	13,9	9,3

Acts of political protest usually occur in a community context and prospective protesters can be encouraged or discouraged from taking part in protest action depending on whether the communities in which they operate are sympathetic or hostile to their behaviour. The "parameters of licence" to protest extended in the Afrikaner community to its members are very restrictive. There is no tradition of protest in the

Afrikaner political culture and the recent and most visible forms of political protest action in South Africa has almost invariably emanated from sources that are hostile to the system of government and the racial ordering of society. Hence Afrikaners feel that protest action is directed at them rather than spent in the advancement of a good cause.

On the other hand peaceful forms of protest have much wider support among members of the English-speaking community at large and certainly have become a stable means of displaying displeasure at the policies and practices of the government at the English-speaking campuses. As far as the white communities go, universities are the most prominent and visible sites where protest actions against public authority are staged. The "parameters of licence" for protest action are quite extensive in the proximate community of English-speaking universities. These factors can help to explain the different levels of protest potential we observe in the Afrikaans and English-speaking students in our sample.

They also support the contention that the level of protest potential observed among students is likely to be the highest of any sub-community amongst whites and certainly higher than the white community at large. Given the relatively low protest potential levels observed among students, the extent to which the ANC and the MDM can rely on support from members of the white community in campaigns of protest against the state is likely to be negligible.

This contention is supported by the levels of repression potential observed in the student sample. Tilly has suggested that a state facing a revolutionary challenge can be rendered more vulnerable if, everything else being equal, it is inhibited from acting forcefully against its revolutionary challengers by resistance from its own support groups. The extent to which the South African state can be rendered vulnerable on this score can be indicated by the support it receives in the white community when it acts to suppress public protest emerging from the black communities in South Africa.

(This index included items such as police breaking up peaceful demonstrations, courts imposing sentences on people taking part in boycotts, arrest of people who take part in protest marches, prison sentences for refusing to do military service, the summary dismissal of workers who take part in stay-aways, the expulsion of students taking part in university boycotts, police firing at demonstrators who damage property and the defence force breaking illegal strikes.)

The depth of support for repressive action by the state in the Afrikaner community and thus in the ranks of the two parties to the

Table 11.10: Repression potential among white students

	Afrikaans %	English %	Total %
No repression potential	5,1	34,4	17,4
Some repression potential	6,6	16,6	10,6
High repression potential	28,9	31,6	30,0
Very high repression potential	59,3	17,8	41,9

right of the political spectrum is evident. But even the profile exhibited by the left of centre Democratic Party differs sharply from the attitudes of supporters of the UDF. Almost 60% of the members of the Democratic Party approve of one or other form of repressive action against the type of protest initiatives typically promoted by organisations like the ANC and the MDM. (To enter the scale at its lowest point respondents had to approve of police firing at demonstrators who damage property. The high point was established by endorsement for police breaking up peaceful demonstrations.)

On this evidence, combined with the evidence on levels of protest potential, the state need have no inordinate fear that repression will be seriously resisted or discredited by strategic sections within the white community.

Attitudes towards a prospective black government

For whites in South Africa, accepting a majority rule model of government means accepting that black political elites will wield significant and probably decisive political power in the polity in future. To tap feelings towards this prospect, we used an index comprising 16 items which required respondents to evaluate the conditions and quality of life they expected to experience under a white-controlled and black-controlled government respectively. On this index, 54% of the students in our sample gave a white-controlled government a high to very high score. 24,6% were neutral and 21,6% gave it a low to very low score. A prospective black-controlled government was given a low to very low score by 59,5% of the students, with 28,9% neutral and 11,5% in the high to very high categories.

The distributions by language group and political party affiliation were as follows:

Table 11.11: White-controlled government

	Afrikaans %	English %
High score	72,2	28,2
Neutral	17,9	33,0
Low score	9,8	38,8

This disposition can be illustrated by a breakdown of the average scores the supporters of the different parties gave a black-controlled and a white-controlled government respectively. On a scale of 1 (low) to 7 (high) for each of these two indices, the following results were obtained.

Table 11.12: Average scores

Average score for a black-controlled government			
CP	NP	DP	UDF
2,0	3,0	3,7	4,9
Average score for a white-controlled government			
CP	NP	DP	UDF
5,6	5,2	3,7	2,5
Average difference: (+) denotes white and (—) denotes black preference			
CP	NP	DP	UDF
+3,6	+2,2	0	—2,4

From the items in the two indices measuring an evaluation of black- and white-controlled governments respectively, we constructed three sub-variables — one measuring a dimension of security, the second a dimension of efficiency and the third a dimension of justness.

We found consistently higher correlations between the variable tapping a dimension of justness and the variables measuring attachment to the state than we did for the efficiency and security variables when they were correlated with the indeces tapping attachment to the state.

We illustrate with a couple of examples. The table shows Pearsons correlation coefficients between the variables just, secure and efficient with the variables measuring attachment to the state.

Table 11.13: A white government scored for justness, security, efficiency

	Just	Secure	Efficient
Attachment to:			
The "core" state	r = ,66	r = ,62	r = ,52
The security establishment	r = ,68	r = ,62	r = ,53
Sympathy with the ANC	r = ,62	r = ,57	r = ,62
Placement on a left-right party continuum	r = ,70	r = ,65	r = ,62

From this we conclude that those who endorse a white government appreciate it not only for its value as a harbour of white security but even more because they deem it good, honest, free and just (the four values from which this variable was built up).

By using factor analysis techniques we were able to establish a single factor or summary variable which can explain all the variables in our study that in some way or other tapped attitudes associated with trust in government, the legitimacy of the system of government, the image of the South African state and its major opponents, racial attitudes and policy preferences. We take this factor to be loyalty towards and support for the state.

We built a composite index of state support by adding together all the variables with a bearing on this factor. They included: protest potential, repression potential, support for the left, support for the intra-system opposition, affect towards state institutions and the security establishment, a left-right continuum built on party identification, an assessment of the legitimacy of the regime and its representatives and the attitudes towards black- and white-controlled government we referred to above. This variable can serve as a summary of attachment to the state. The total distribution was as follows:

Table 11.14: Attachment to the state (total sample of white students)

Very strongly attached	13,7%
Strongly attached	45,3%
Attached	23,9%
Becoming detached	12,3%
Already detached	4,9%

The distribution by language group looked like this:

Table 11.15: Attachment to the state by language group

	Afrikaans %	English %
Very strongly attached	22,4	3,5
Strongly attached	58,9	26,0
Attached	13,9	36,2
Becoming detached	3,7	23,7
Already detached	1,1	10,5

Conclusion

The Afrikaner formations assembling in the fold of the right-wing parties might be disaffected with the National Party, its leadership and with its principal reformist policies but we can find no evidence that this has translated into a commensurate alienation from any of the institutions that represent the authority of the state in civil society.

Differences in the response patterns on all of the indices that probed for affective attachment to the *state* could be accounted for by the language group divisions in the white community rather than as emanating from the political fractures inside the Afrikaner community. The one exception is responses on an index that probed for feelings of trust in *government* — a factor that could pick up the disaffection of right-wing supporters for the National Party incumbents to government rather than for the structure of government itself. On this index the recorded levels of distrust almost matched those found amongst English-speakers and supporters of the Democratic Party. (The trust index included items that required respondents to evaluate government incumbents for their honesty, wastefulness, efficiency and trustworthiness.)

The major lines of division run between the language groups in the white community and between those in the white community who operate within the polity and those who align with the forces challenging the state from beyond the bounds of the polity. As our summary variable indicates, Afrikaners assembled in both the National Party and the Conservative Party are at the core of the white fortress and in control of its major instruments of coercion and administration. The English-speaking section showed serious signs of disaffection during the final years of the Botha administration, but this trend has been reversed as the new De Klerk administration legalised the liberation movements and started to talk to the ANC.

The outside perimeter of the Afrikaner "laager" is opening up only in a modest way. Amongst the students in our sample, representative of the younger generation, support for the liberal Democratic Party has increased about threefold in the last couple of years. Very small, but symbolically very significant and previously unthinkable, defections from the Afrikaner community to the ranks of the liberation movements have occurred. An Afrikaans newspaper pleading the cause of the liberation movements has been on the streets for almost a year now and "negotiations with the ANC" is a fashionable phrase uttered approvingly even in the ranks of students supporting the National Party.

When weighed against the hold of the state on the white community, the fractures and divisions within the white ruling group are weak political forces. The Afrikaner political satirist Pieter-Dirk Uys has suggested that in South Africa "paradise is closing down". Closing down, maybe, but not cracking up.

ENDNOTES

1. The sampling frame was established on the basis of all intra-murally enrolled white students at universities in South Africa where the majority of students are white. (Only the University of Port Elizabeth refused their co-operation.) A 10% simple random sample was drawn. A postal survey was conducted and three waves of questionnaires were sent out at fortnightly intervals. Of the 8 747 questionnaires sent out with each wave, 4 321 usable ones (just under 50%) were ultimately returned. As the percentage returns varied between universities, a weighing procedure was introduced in the statistical processing of the data so as to put all the universities on a par. We found that the distribution of the known demographic and associational characteristics of the populations at and across the different universities were very accurately reflected in the sample.

2. Theda Skocpol: *States and Social Revolutions*, Cambridge, Cambridge University Press, 1985, p32.

3. Samuel Huntington: "Reform and stability in a modernising multi-ethnic society", *Politikon*, Vol 8, 1981, p11.

4. The indices of Tables 11.2 to 11.5 reflect feelings towards the state, the co-optation institutions of the state and the security establishment, which were constructed by performing a factor analysis procedure on responses to some 35 of the major political organisations and institutions that operate in the political landscape of South Africa. We used the so-called "feeling thermometer" as the stimulating device and required our respondents to indicate feelings of sympathy/antipathy on a five-point scale. The indices were constructed on the basis of the factor loadings yielded by the factor analysis. The "core" state is a composite index (in the additive Likert format) of responses to the following items: the civil service, the state president, the defence force, the police, South African Television (SATV), the National Party, the House of Assembly and the courts of law. The co-optation structures included: homeland

leaders, Inkatha, the House of Delegates and the House of Representatives. The security establishment comprised the security police, the South African Defence Force and the South African Police.

5.   The options were circumscribed as follows:

Partition: A geographically divided South Africa, one part of which is controlled by whites and the other by blacks, with the two parts totally independent of each other.

White control: A system with a white-controlled government on the central level, in which "non-whites" are either incorporated in a subservient role or exercise their political rights in their own subservient institutions.

Racial/ethnic federation: A system in which racial and/or ethnic groups are represented as groups in the central government, with no single group having a final say.

Geographical federation: A federal system dividing South Africa into a number of multi-racial geographical regions represented on an equal level in the central government, in which the rights and interests of groups are protected.

Unitary state: A unitary state in which all South Africans take part in the same electoral process and government is controlled by the majority, with no special protection for groups.

Working-class state: A system in which class distinction is the prime criterion and the working class controls the government.

12 The radical challenge

NEVILLE ALEXANDER

In her celebrated attack at the turn of the 19th century on the revisionist doctrine of Eduard Bernstein, Rosa Luxemburg formulated with crystal clarity the position which I take up in this chapter:

> Can the social democracy be against reforms? Can we counterpose the social revolution, the transformation of the existing order, our final goal, to social reforms? Certainly not. The daily struggle for reforms, for the amelioration of the condition of the workers within the framework of the existing social order, and for democratic institutions, offers to the social democracy the only means of engaging in the proletarian class war and working in the direction of the final goal — the conquest of political power and the suppression of wage labour. Between social reforms and revolution there exists for the social democracy an indissoluble tie. The struggle for reforms is its means; the social revolution, its aim.[1]

In essence, the difference in approach to this question of strategy between revolutionary socialism on the one hand and certain strands of bourgeois radicalism and modern social-democratic tendencies on the other hand, is the belief of the former that in order to bring about a more equitable mode of distribution of the social product, it is essential that the mode of production be radically changed. The latter are content to accept that a more "just" society can be brought about by merely changing the mode of distribution within the framework of the capitalist mode of production. Revolutionary socialism takes as its point of departure the unavoidable need to restructure the capitalist state in accordance with the interests of the new ruling class, that is, the victorious working class. Consequently, it postulates that somewhere along the line of march of the forces of liberation there will be a more or less violent clash between those who wish to conserve the existing capitalist mode of production and those who wish to displace it with a new (socialist) mode of production.

In South Africa, there is a debate on the radical left about whether the struggle against apartheid entails one or two revolutions. Some, notably the orthodox Communist Party, argue that the South African revolution will go through two stages, the first a relatively violent one against the apartheid system of racial discrimination, the second a relatively peaceful one for the eradication of capitalism. This is essentially a debate about the class leadership of the anti-apartheid struggle. If the hegemonic leadership represents middle-class interests and aspirations, two stages are inevitable, not so if the leadership is rooted in working-class interests. Consequently, this debate boils down to one's assessment of whether there is any objective basis for postulating the probability of "working-class leadership of the struggle against apartheid". The corollary to this is obviously how one understands the notion of "working-class leadership".

This is all the more relevant in the context of developments in Eastern Europe and in the People's Republic of China where most, if not all of the holy cows of "socialism" are being slaughtered.

It may well be asked whether a concept such as "revolutionary socialism" can still have relevance in the South African struggle, especially in view of the fact that, for theoretical and more transient political reasons, some currents within that struggle question the proposition which holds that the immediate goal of the struggle is the realisation of a "socialist society", however defined.

On this question, it is essential to make two points unequivocally clear. Firstly, whatever one's assessment of the "Gorbachev revolution", it is simply a logical fallacy to aver that because the processes which have been initiated amount to a fundamental revision of many of the practices and theories of bureaucratic socialism à la Stalin, these processes validate the bourgeois-liberal critique of socialist and Marxist theory. It is unnecessary to dig any further into this particular complex; suffice it to say that the hollowness of the syllogistic argument from which this inference is derived is self-evident. Secondly, their experience of "actually existing capitalism" in South Africa objectively predisposes most black people there to the negation of the capitalist system, whatever the particular forms in which this aspiration is mediated in their subjective consciousness.

Liberal free-marketeers can shout until they are blue in the face that what we are dealing with in South Africa is not capitalism but some degenerated form thereof, but for black people generally, and black workers in particular, the present apartheid modality of that system is capitalism as they know it. This situation is no different in principle from that in the so-called socialist societies. Whatever the terminological and conceptual intricacies in which Marxist scholars and activists engage concerning the notion of "socialism", for the people of Eastern Europe, the societies in which they live portray socialism as they know it.

However, the point of this comparison is simply to suggest that those who are bold enough to propagate a socialist alternative in South Africa today are clearly obliged to go back to the drawing-board in order to think through and research from first principles the basic framework of a socialist world. Moreover, they have to understand that in practice it is not the notion of a socialist utopia or even of "scientific socialism" that will guide their day-to-day activities in the epoch of the transition from capitalism to socialism. It is rather some such notion as "feasible socialism" (*pace* Alec Nove) that in fact influences what socialist activists of one kind or another actually do in the world today.

This point has been reinforced by the return on the part of Gorbachev and his circle to the acceptance of the pristine Leninist-Trotskyist *a priori* that the world economy is an integral whole and that "socialism in one country" is simply a vicious myth. At present, the world economy continues to be dominated by the capitalist mode of production even though this obviously does not mean that in capitalist and transitional societies elements of a socialist future cannot be explored and consolidated. Analogies are never accurate but good analogies are always illuminating and I believe that we need to see the present post- or non-capitalist societies in the "socialist" half of the world as similar to the Italian city states in the period of the transition from feudalism to capitalism in Europe. Moreover, every new thrust towards a socialist future, wherever it may originate, imposes on the conscious elements in such a society the task of systematising the experience of the masses of the people — who are the actors that matter — in ways that define the content of socialism and point out the road ahead.

Reform in South Africa

There is no need in a short chapter such as this to enter into any detailed analysis of the universally acknowledged current reform process that was set afoot in South Africa somewhere between 1976 and 1978. Whether or not one considers such reforms as the National Party has introduced since then to be "cosmetic", there is no doubt at all that:

> The point for the political establishment in initiating reforms
> . . . is to alter established priorities and structures that would
> endanger the basis of its own power. Since reform is a
> purposeful attempt of a political leadership to remove a danger
> to itself, it is to be expected that it will not entail new policies,
> but more stop-gap measures accompanied by repression . . .[2]

From the point of view of the ruling capitalist class in South Africa, there are three overlapping options out of the organic crisis. Of these, only two can be labelled "reformist". The first, or counter-revolutionary option is propagated by both the Conservative Party and its flat-earthist allies on the right in the extreme form of a South Africa partitioned for the benefit of the white minority, as well as by the conservative elements within the National Party. In this connection, it is pertinent to note that:

> The official position of the National Party government
> remains the insistence that, in PW Botha's words: "As far as I
> am concerned, I am not considering even to discuss the
> possibility of black majority government in South Africa."
> Preventing this outcome remains, in essence, the purpose of
> "counter-revolution" in South Africa.[3]

The second option is that of the radical liberals who are to be found among parliamentary as well as extra-parliamentary currents in South Africa.

> They believe that an alternative capitalist-orientated power
> bloc has to be established, because they do not think that the
> National Party . . . can ever gain the support of even a small
> percentage of black people . . . For the realisation of this grand
> strategy against Afrikaner nationalism, which has loyally
> carried the flag of capitalism in Southern Africa up to the
> present . . ., they hope to be able to harness that other great
> social force in South African society, that is, African
> nationalism.[4]

The centrepiece of the alternative bourgeois coalition, in this view, is to be constituted by the African National Congress, the traditional custodian of African nationalism in South Africa. Those who are promoting this option are to be found in most organisations to the left of the National Party as well as among a minority of individuals in that party itself. The central insight of these people is that:

> . . . the crux of South Africa's political problems . . . (is) the
> absence of a legitimate constitution capable of meeting the

demands of the majority while addressing the fears of the minority. It is largely because these two imperatives are widely regarded as irreconcilable that South Africa's conflict continues ... (Hence we have to ask): How do we reach the point at which meaningful constitutional negotiations can take place?[5]

In practice, those who are pursuing this option include people who are committed to classical, one-person-one-vote bourgeois democratic blueprints as well as others who tend towards consociational-democratic solutions.

In the centre between these "extremes", we have the present pragmatic option typified by the nexus of De Klerk, Pik Botha, Malan and Du Plessis in the National Party. This policy of a slightly jazzed-up "co-optive domination" sees the National Party as being the main vehicle for managing the transition from the present apartheid to some multiracial, "post-apartheid" set-up, one which is open-ended in so far as it will permit the inclusion of any element tolerable to the social base of the National Party but which does not threaten the ultimate control of power by that party. This approach is aptly described in the words of Gustafson:

> Any reform amounts to a deliberate attempt to consolidate or rescue what the leaders consider essential whilst offering up the non-essential for sacrifice, hoping that the forces of change that they have let loose can be kept under control.[6]

The regime cannot, however, escape the dilemma of all reformist pre-revolutionary governments, that is, that all reforms demonstrate the illusory character of reformism. In this respect, a poignant parallelism can be identified between the modalities of the present custodians of the apartheid state and the bureaucratic centralist regimes of Eastern Europe. In an exceptionally illuminating article, Petr Uhl, a member of the Czech opposition movement, writes about the contradiction between the attempts of the bureaucracy to maintain the status quo and the need for social change and notes that:

> The institutions of bureaucratic centralism are unreformable, but minor improvements within the framework of the system are important because they encourage the development of a critical spirit, a mood of opposition, and embryonic structures independent of the state. Reforms in themselves have limitations that are well-known: they always stop — or are repressed by terror — when they touch on fundamental solutions to social contradictions. In this sense every attempt at reform has a revolutionary aspect to it, for it reveals the illusory nature of

reform and strengthens the growth of a revolutionary consciousness.[7]

Revolution in South Africa

The conquest of power by means involving a degree of revolutionary violence is predicated on the objective fact of antagonistic class contradictions. That is to say, the fact that the promotion of the interests of one class or of a combination of classes (or fractions of classes) *necessarily* implies the erosion or undermining of the socioeconomic base of another class or combination of classes. This *a priori* of all revolutionary theory remains valid in South Africa unless the ruling group(s) there are going to behave differently from ruling classes throughout the history of the world. In Europe, it hardly needs restating, almost all major or fundamental advances in the political and social spheres have come about ultimately by means of more or less violent revolution.

In South Africa, Marxist and some Africanist groups have promoted the strategy of violent revolution at the rhetorical and theoretical levels since the 1920s. In 1960, most credible political organisations were forced by the National Party to resort to "armed struggle" which, needless to say, is not to be equated with "revolution". Armed struggle, however, has widened the catchment area for the recruitment of revolutionaries. To such an extent, indeed, that the pre-1960 passive-resistance, non-violence, reformist mould of the entire national liberation movement can be said to have been irrevocably broken. This is the reason why the present orchestrated moves to "create a climate of expectation" have become necessary. In a paradox of no mean magnitude, a downtrodden people who for centuries have been relegated to slave status, acquiring thus an instinctive disposition to petitionist and reformist approaches, now has to be persuaded that negotiations and talks may actually work.

Most black people in South Africa have come to believe that talking to the South African government is a waste of time: they have come to understand the fundamental contradiction (at all levels) between a non-racial and a racist approach to the problems of this country. There is, again, hardly any need to spell out the details of the situation. Steven Friedman, an insightful commentator by any standard, has very recently remarked how vulnerable those are who are willing to consider "negotiating" with the present government. Noting that the Africanists and the Black Consciousness currents in South Africa are opposed to negotiations, unlike the "Mass Democratic Movement" and the Charterists, he goes on to ask about the BC camp:

> Do its views matter? Some see it as a small splinter group which can afford to take extreme positions because it has little support and which can be largely ignored. That assessment may well be a serious mistake. The BC groups don't have as much of a mass following as their rivals. But their ideas do have significant support — and even more potential support. That potential is most likely to be realised if momentum for a settlement grows . . . Nor do its ideas only enjoy support among its own members — they also have a sizeable following within the ANC and MDM. A fair number of MDM activists learned their politics in the BC camp or on the Cape Flats, a stronghold of the tradition which rejects all compromises and "collaboration" with the system. And, regardless of where activists learned politics, opposition to negotiation and compromise strikes strong chords within the entire black resistance movement, including the MDM.[8]

The conquest of state power ultimately revolves around whether the revolutionary forces can destroy or defeat the repressive organs of the reactionary state, in particular the standing army of that state, or persuade them to remain neutral or to go over to the revolutionaries. Judged by this criterion, the possibility of a revolutionary overthrow of the apartheid state is obviously extremely remote. The military-technical conditions for the conquest of state power by the forces of radical change in South Africa can be spelled out clearly. Four conditions are indispensable.

1. The South African Defence Force and other repressive organs have got to become "blacker", that is, the class composition of the army has to change until the critical mass is attained where it will become responsive to the revolutionary people. At present, the SADF is not a popular army in the sense of being drawn from all sectors of the population. It is, in fact, insulated against the people by racism and is, thus, in more ways than one, the shield of the ruling class against its class enemies. The maturation of this process is bound to take many years since the rulers are as much aware of its significance as are the revolutionaries.

   The South African ruling class is caught in a contradiction. To rely primarily upon white soldiers and policemen would be both to suffer endemic manpower shortages and, if conscription were tightened even further, to increase the already serious rate of white emigration, especially of professional and managerial employees. But to look instead towards armed forces composed largely of black mercenaries would be to place their future on troops who might easily switch sides — like the black troops crucial to the Rhodesian settler regime's counter-insurgency war against Zimbabwean guerillas, who rapidly changed allegiances after the ZANU-PF election victory in March 1980. Pretoria is likely, here as so often, to vacillate, relying on a core of white troops but recruiting more blacks.[9]

Recent examples of the mutiny-proneness of black soldiers and policemen (Rockman, Holomisa, South West African Territory Force) point to the vulnerability that will come into the security apparatus as its class/colour composition changes. Although it would be possible to reduce this vulnerability by ensuring that the security forces become one of the main beneficiaries of the system, as long as racial discrimination continues to characterise South African society, this potential vulnerability will remain. It is, in any case, a significant datum that probably more black men and women are being and have been taught the basics of using lethal weaponry than are being trained by the ANC, the PAC and other South African insurgent groups.

2.   Although all South African revolutionaries have come to accept that the classical guerrilla strategy pioneered by the Chinese Communist Party does not, by itself, hold any possibility of success, it is also clear that the escalation of urban guerrilla attacks and sabotage as well as industrial sabotage is an important aspect of the "continuation of politics by other means". The guerrilla strategy initiated in the 1960s has acquired a logic of its own and, like the developments in Northern Ireland, forms a kind of backdrop to the political stage on which every new generation of South African activists has to make its contribution. From the point of view of the conquest of power, the auxiliary role of the guerrillas is of inestimable importance in so far as they will tend to tie down thousands of soldiers and police in patrol duty and in related functions.

3.   The ability of the armed wings of the liberation movement to arm the working masses must increase. No militia tradition exists among black South Africans. However, spontaneous township insurrections in recent years, sometimes fanned by activists and

trained guerrillas, have inaugurated a pattern which could soon develop the force of tradition. From there, it is a short step to the attainment of the situation where thousands and even hundreds of thousands of people will be able (and willing) to use arms against the agents of the state, should conditions necessitate. Once this position arises, the days of racist South Africa would be numbered. It may be pertinent to remark that there are vested interests even within the liberation movement itself geared to retarding, and even preventing, this situation from being realised because of its obvious radicalising potential. Given the number of activists, black and white, who are now able to train people in the use of arms inside South Africa, this particular development appears to be unstoppable, however.

4. South Africa's neighbours have to become stronger economically and militarily. Abstracting from all other implications of the analogy, what I am suggesting is that countries such as Zimbabwe, Angola, Mozambique and, perhaps, Namibia should become *mutatis mutandis*, like Syria, Iraq, Egypt, and so on, around a South African Israel. This would mean that in any limited international conflict involving South Africa and the African states, the armies of these countries would be able to checkmate the SADF, not unlike the manner in which FAPLA (Forcas Armadas Popular para Libertacao d'Angola) and Cuban forces stopped it at Cuito Cuanavale. The ensuing power vacuum would without any doubt rapidly be filled by the organised black working class.

Intersecting strategies

The maturation of these four conditions would normally take a long time but the process can be expected to be drawn out even longer because of the fact that the South African authorities and their strategists understand this better, almost, than anyone else. While unforeseeable and unforeseen developments can easily telescope events in the manner of, say the First World War in the case of the Russian Revolution or the Portuguese army revolt in the case of the Mozambican and Angolan revolutions, it is clear that a revolutionary strategy in South Africa today is *necessarily* a long-term one. It involves, among many other things, creating the socio-cultural and the organisational conditions that will ensure the reconstruction of society and the state in accordance with the interests of the new ruling class(es) after victory.

This is why the phrase "building tomorrow today" has come to encompass in a nutshell the fundamental strategy of all revolutionary forces in South Africa. This is why the strategy of the "long march" resolves the apparent contradiction between the struggle for social reforms, for the improvement of the condition of the working class on the one hand, and the struggle for the conquest of state power on the other hand. This is why, often, the day-to-day tactics of the revolutionaries — who caution against short-term illusions and compromisist agendas such as indefinite school boycotts, long-term consumer boycotts, vaguely defined defiance campaigns to enhance the bargaining power of those who wish to "negotiate" — seem to be "conservative" and have even been denounced by would-be "revolutionaries" of the short-term variety as "reactionary" and "counter-revolutionary".

But does this mean that the struggle for social reforms as an alternative to the negotiation option is similar to the strategy of reformist liberalism and social democracy in South Africa? The answer is clearly in the negative. The ends or goals of these two trajectories are different and their starting points are different because they represent different social analyses based on different class interests even though their means appear to be similar at points of intersection in the socio-political domain.

Before we look at a specific example, it might be useful to establish the point that *any* reformist option in South Africa will only be successful if it is historically possible to disentangle racism and capitalism, or, to put it differently, if a non-racial capitalism is attainable under the historical conditions in which the struggle is being fought in South Africa. There are at least two important reasons for believing that this is unlikely.

The first reason derives from the modalities of the capitalist mode of production. Under the historical conditions of the South African social formation, that mode of production, left to itself, as it were, will continue to reproduce *social* inequality predominantly as *racial* inequality. However that inequality is labelled: "apartheid", "segregation", "separate development", or anything else, the overwhelming majority of black people will continue to believe that the capitalist system is prejudiced against them. Hence, they will continue to fight against it for that radical and rapid redistribution of wealth which would be tantamount to a revolutionary reconstruction of the entire society.

This is the dilemma of the liberals in the ruling caste, and this is the source of the second reason why the real alternative in South Africa is either socialist revolution or some kind of "Lebanonisation" of the society. For, as has been pointed out repeatedly, the supposed contradiction between capital and the state in South Africa which is supposed to derive from the dysfunctionality of racism in regard to capitalist development has only occasionally and episodically become manifest. While it is true that racism and capitalism are contingently linked in South Africa, as elsewhere,[10] it is also true that the particular entanglements of South African history in the post-war period leave "capital" no option but to come down on the side of the "state" in the final analysis and to take up position against any alternative based on majority rule even if it is clearly defined within a liberal capitalist framework. The analysis of Callinicos in this regard is, in my view, accurate and compelling:

> While there are differences between big business and the National Party government . . ., they are agreed in seeking for political devices which would permit the political incorporation of the black middle class while preserving effective white control over the state.[11]

He cites Anthony Sampson as saying about the Lusaka safari led by Gavin Relly of Anglo-American in 1985:

> (The businessmen) knew that they would not run their factories and mines for long without the consent of the black workers, and that black political ambitions could not be suppressed indefinitely. Many even agreed that there could even be a black government in ten years time. But they could not visualise any likely transition between now and then. They were prepared to offer blacks almost anything except the one thing they demanded: one man, one vote. Most found it impossible to contemplate majority rule or to deal with the blacks who were likely to command it.[12]

Callinicos infers irrefutably that:

> The central difficulty lies in the transition to black majority rule. The ANC could perfectly well provide capital with a bourgeois black government which would guarantee the conditions of continued capital accumulation . . . But capital also needs to be able to rely on the existence, during the transition to majority rule, of a state apparatus capable of defending bourgeois property relations against the black working class. Such a state apparatus exists at present — above all in the formidable might of the SADF and the SAP. But the cohesion of the existing state machine is inseparably interwoven with the

white population and the material privileges they derive from the prevailing system of racial domination.[13]

All this means that in the political domain, no strategic or organic alliances are possible or viable as between reformists and revolutionaries in South Africa. The rapidity with which the flirtation between "parliamentary" and some "extra-parliamentary" groups opposed to the National Party government's strategy has come to an end demonstrates the point at a relatively trivial level. In the words of Milazi:

On the one hand, are the ruling party's reforms and (the liberal) alliance's paradigm of transformation and evolution, suggesting peaceful incremental growth towards a "post-apartheid" order in South Africa, and, on the other hand, the world view of the liberation movement largely dominated by a revolutionary paradigm in terms of which an immediate and total demise of the prevailing order is required.[14]

He concludes that

Liberal leadership cannot effectively be part of such a (national liberation) struggle — for it cannot practise what it *has* to preach. Its liberalism — in radical parlance — is consistent with the hegemony of the bourgeoisie ... Liberalism is (and remains) a rhetoric — for the simple reason that the bourgeoisie as a class cannot produce the conditions of economic democracy which are the necessary condition of the realisation of its own rhetoric. Therefore, in its current state of ideological and material interests, the liberal alliance is unable (or unwilling) to assume the role of revolutionary elites to build coalitions of classes and oppressed groups.[15]

Of course, this proposition does not exclude the possibility of episodic tactical alliances between liberals and radicals in the political domain.

If, however, viable strategic political alliances between liberal or social democrats and revolutionary democrats are unlikely to materialise, this does not mean that all co-operation is impossible. Indeed, the very fact that both parties are committed to social reforms in spite of having quite different agendas holds within itself the possibility of intersecting and mutually reinforcing strategies. By way of demonstrating this point as well as some of the problems inherent in the thought, let me quote at length from a summary of a 1989 workshop organised by the Human Rights Trust in Port Elizabeth. This workshop, like the conference at which this paper was delivered, set out to examine "Strategies to end apartheid"!

Speaker 1: I believe that to ask black organisations to form alliances, even tactical alliances, with parliamentarians is not on.

Speaker 2: The question becomes: we have white people committed to many values that are shared by those in extra-parliamentary politics — non-racialism, democracy, universal franchise, the rule of law, a basic set of universal human rights, and other commonly agreed issues — and they are wanting to use the traditional white political culture, which includes elections and parliament as central features, to create a platform for these issues. Now what they want from the extra-parliamentary movements is, ideally, help and co-operation in joint campaigns in this regard, or, less ideally, non-interference in their/the white's campaigns. But what you are telling us is that we could quite well expect more of the 1987 package, which we would term undermining . . .

Speaker 1: I don't see any chance of any alliance which is going to give a mandate to people to enter racial political structures to demand, as it were, the franchise for us. Obviously we'd never say, "don't demand universal franchise", though.

Speaker 2: What you are saying is, of course you can go and demand universal franchise, don't wait for us, and yes we demand it also. But don't do it by running for an election for a racially defined parliament, no matter how successful that campaign might be to win universal franchise.

Speaker 1: I am not prepared to use a racially defined platform, but I will not deny you that right if you decide you can do good things there. But I reserve the right to criticise you if I see you in practice doing things that, while maybe popular with a white electorate, cut across my principles or policy.

Speaker 3: You cannot expect those people outside of parliament to put all their eggs into the parliamentary basket — but we accept institutions as given, and should not hinder you in using them to break down racial politics.

Speaker 4: We are told to organise in our white constituency — but our success here depends on our ability to deliver non-racial politics. Thus we cannot succeed without the help of extra-parliamentary movements!

Speaker 5: My position differs from Speaker 1 somewhat. Let's look at what has happened recently. Some parliamentarians have left parliament for extra-parliamentary groups. This has bridged the gap somewhat. The safaris to Lusaka have also. For white parliamentarians to change attitudes to participation, they must follow up the examples of parliamentarians who go to places where blacks are shot, and who face the firing line with them. If this interaction is opened up, the oppressed will feel that they are being supported and they will support in turn.

Speaker 1: Parliamentary politics in an apartheid state is the politics of dissension concerning the status quo, or a slightly altered status quo. It in no way addresses the revolution that is required, and is inevitable, to convert our society into a post-apartheid society. Now the people in these racial structures can do some good — they are attacking crude racism, and are doing things that from my point of view, will have to be supported. But there is a limit beyond which an officer elected by a white electorate can't go, and expect to be re-elected. Thus one mustn't delude oneself that you are going to get the majority of blacks into some sort of alliance that gives a vicarious power to white MPs or city councillors. I don't think that can happen, because voters want benefits from the MPs from a system that continues to operate on racial repression of the people who are now expected to support people to election in that system.[16]

The issue of alliances has got lost on the one question of extra-parliamentary people somehow giving approval to some white politicans to participate. But there is another type of tactical alliance that I am talking about that could happen without any parties to the alliance ever talking to each other, let alone agreeing with each other. They could just be certain groups working to the same objective. An example would be the community organisations that grew up in 1984–86. They were obviously engines for change, and there was also evidence of parliamentary groups (in the Progressive Federal Party) trying to make space for them to operate, and trying to get legitimacy for them under the umbrella of the "system" so that they would not be crushed. Each parliamentary group could help the extra-parliamentary in the way I have outlined.

But then the extra-parliamentary grouping got projected as a vehicle for the fairly imminent overthrowing of the state, and then it was impossible for a parliamentary group (or any other) to provide it with support, and the organisation was on its own. Now, to put in a crude question, "if you do not have the muscle to overthrow the state, is there much point in behaving as though you do?" The fate of the extra-parliamentary groupings seems to imply that there is no point in so behaving.

Clearly the main area of intersection is in the domain of civil society, which the Eastern Europe opposition refers to as the "parallel polis" or "alternative community". This is the terrain on which counter-hegemonic practices and institutions are being established and experimented with. It is the terrain of the cultural revolution which, in these highly repressive societies, necessarily precedes the political and social revolutions. Here the values and institutions of the ruling group and, more generally, of the ruling class, are put into question both through the spontaneous rejection by the working people in their hundreds of thousands and through the conscious and systematic intervention of core groups of activists variously organised in accordance with the peculiar histories of the resistance in their respective countries. This alternative community constitutes the base on which the post-revolutionary (or, more neutrally, the post-apartheid) society will be built.

In South Africa, as is well known, such an alternative community emerged in the period 1976–80 for the first time, although the indispensable preliminary work for its emergence was done by the Black Consciousness Movement in the preceding five years. Since then, it has grown, despite brutal repression, so that today the country is criss-crossed by a network of "people's organisations" or community-based groups ranging from independent trade unions and churches with hundreds of thousands of paid-up members to small service organisations or two or three activists working against racism and sexism in education. In all these organisations, the vision of a future non-racial, democratic South Africa is being coloured in, as it were, with crayons representing the infinity of experience and dreams of an entire people struggling to negate the nightmarish reality of despotic racist rule.

It is in the promotion of these projects, large and small, that democrats can co-operate, even if not without friction. The critique of racism, sexism, authoritarianism and undemocratic practices is common to all democrats in South Africa. Projects such as some of the less romantic and high-flying ventures into "people's education", the National Language Project and numerous literacy and pre-school projects with national scope — to mention only a few — represent the best areas of co-operation. They represent, moreover, opportunities for intervention and contribution at a profound level. All of them are indispensable if the ideas of the liberation movement are to become hegemonic. They are necessarily long-term projects of a cultural-political kind which presuppose the possibility of realising a single South African nation organised in a democratic, non-racial society where there is a maximum of self-organisation of the people.

It is true, of course, that there will be contradictions arising from fundamentally different approaches to the question of social reform. Thus, for instance, liberal democrats may tend to see the alternative community, especially when it has not yet assumed a mass character, as a useful safety-valve for letting off the pent-up revolutionary energies of "the masses". For the revolutionary democrat, on the other hand, it may represent the prefiguration of a future in which the working people will take as much of their destiny into their own hands as they possibly can. Liberals may, for example, see the establishment of co-operatives as a transitional mechanism for encouraging the entrepreneurial spirit among downtrodden but resurgent people, whereas socialists may see them as hybrids or as what Luxemburg calls "small units of socialised production within capitalist exchange". Certainly, these differences will lead to tension and contradiction but they also permit a large measure of co-operation. Similarly, in the communicative aspects of language and literacy projects, both of which have important immediate economic spin-off effects, there is a large measure of overlapping interest among all democrats and anti-racists.

The list of possible candidates for co-operative efforts is a long one provided it is clear at all times that at the end of the day we are not going to be able to avoid the really difficult questions. We would not be human if we did not hope that in the course of co-operation in these various spheres our different realities would be corrected and enriched by the process of co-operation. On the other hand, reality is a product of our particular conceptual universes and one can only begin to live in and experience the reality of the other if one accepts his/her conceptual universe.

This is the Great Wall of China that separates bourgeois democracy from socialist democracy, whatever the superficial resemblance. It is, in fact, the difference between the representative democracy of the classical parliamentary type and the direct democracy of the type crudely practised in independent trade unions and, perhaps, in a more advanced form in communes and in pre-Stalinist soviets or workers' councils. It is the difference between a belief in mere legislative reform of the system and the need to transform the system itself. This is the reason why, in the end, the inescapable question of revolutionary violence will separate the spirits. Even 200 years after the French Revolution liberal democrats are not willing to face the fact that made reality of their own vision of the ideal society, the fact, that is, that:

Revolutions do not happen because revolutionaries call for

them, nor are they a result of indoctrinating the masses. They happen when people choose violence and use it to take power from those who hold it. They happen when the broadest strata of people can no longer bear their oppression, when the incompetence of the rulers goes hand in hand with brutality and terror. The role of revolutionaries is to indicate to the masses in revolt the best way to go, which means, among other things, trying to limit revolutionary violence to the smallest necessary degree and consistently opposing brutality and terror, which, even when it is a necessary factor in the revolution's survival, damages the revolutionary process because it tends to degenerate it.[17]

Conclusion

In the final analysis, liberal and social democrats draw an artificial distinction between reform and revolution. If one accepts, as I do, that social revolutions are historical points that demarcate the formal transition from one epoch of production to another, it is clear that reforms are adaptive mechanisms employed by the last revolutionaries and their epigones to consolidate and defend their revolution. Inevitably, therefore, all reformist endeavours are a paradoxical reflection of the threats that are developing to the existing order. Such threats are usually the result of endogenously generated contradictions between opposing class forces, less often also of exogenous causes. In spite of the obvious circularity of my argument, I can do no better than to conclude this paper with the winged words of Rosa Luxemburg:

> From the first appearance of class societies having the class struggle as the essential content of their history, the conquest of political power has been the aim of all rising classes . . . Legislative reform and revolution are not different methods of historic development that can be picked out at pleasure from the counter of history, just as one chooses hot or cold sausages. Legislative reform and revolution are different *factors* in the development of class society. They condition and complement each other, and are at the same time reciprocally exclusive, as are the north and south poles, the bourgeoisie and the proletariat . . . That is why people who pronounce themselves in favour of the method of legislative reform *in place of and in contradiction to* the conquest of political power and social revolution do not really choose a more tranquil, calmer and

slower road to the *same* goal, but a *different* goal. Instead of taking a stand for the establishment of a new society they take a stand for surface modification of the old society.[18]

ENDNOTES

1. R Luxemburg: *Reform or Revolution*, New York, Pathfinder Press, 1978, p8.
2. D Milazi: "South Africa's Democratic/"liberal" Alliance: Implications for the National Liberation Movement", paper delivered in the African Seminar, Centre for African Studies, University of Cape Town, 2 August 1989, unpublished memo.
3. M Swilling: "Reform, Security and White Power in South Africa: Quo vadis?" in Dick Clark (ed): *The Southern Africa Policy Forum*. First Conference March 28-31, 1989, Queenstown, MD, Aspen Institute.
4. African Research Centre (ARC): *The Sanctions Weapon: Summary of the Debate over Sanctions against South Africa*, Cape Town, Buchu Books, 1989.
5. H Zille: "From Conflict to Resolution in South Africa", in Clarke: *The Southern African Policy Forum*, p29.
6. Milazi: "South Africa's Democratic/"liberal" Alliance", p9.
7. P Uhl: "The Alternative Community as Revolutionary Avant-garde", in Vaclav Havel *et al* (eds): *The Power of the Powerless: Citizens against the State in Central Eastern Europe*, London, Hutchinson, 1985, p190.
8. Steven Friedman: "Negotiations: Its Major Divisive Risk", *Weekly Mail*, 1 September 1989.
9. A Callinicos: *South Africa between Reform and Revolution*, London, Bookmarks, 1989, p189.
10. H Wolpe: *Race, Class and the Apartheid State*, London, Unesco, 1988.
11. Callinicos: *South Africa between Reform and Revolution*, p162.
12. Callinicos: *South Africa between Reform and Revolution*, p163.
13. Callinicos: *South Africa between Reform and Revolution*, pp163-4.
14. Milazi: "South Africa's Democratic/"liberal" Alliance", p13.
15. Milazi: "South Africa's Democratic/"liberal" Alliance", p18.
16. Unpublished and unofficial summary of Human Rights Trust workshop entitled "Levers on Apartheid", Port Elizabeth, March 1989.
17. Uhl: "The Alternative Community as Revolutionary Avant-garde", p190.
18. Luxemburg: *Reform or Revolution*, p49150, emphasis in the original.

13 Cohesion and coercion
HERIBERT ADAM

Cultural versus racial nationalism

Many analysts of nationalism have pointed to the Janus face of
nationalism: the enmity for outgroups paralleling amity for the ingroup.
Ferocious hostility goes together with self-sacrifice and altruism. Love
for home, the special sound of language and music, the superior taste of
food, the landscape of childhood, unique roots and histories, are usually
degraded to an ethnocentric rejection of the stranger. Non-ethnics are
frequently excluded from human concerns at best, exploited and
inferiorised as scapegoats at worst.

Afrikaner nationalism shows a similar Janus face. In the battle with
British imperialism, the Boers were supported by progressive human-
kind, including Lenin. Apartheid nationalism stands unanimously
condemned by the world. Here it is useful to introduce the important
distinction between cultural nationalism and racial nationalism. In the
name of cultural nationalism self-determination can be claimed. Cultural
or ethnic nationalism is a legitimate form of self-expression.

However, Afrikaner nationalism has overlaid its cultural achieve-
ments with racial nationalism. It excludes brown Afrikaners from its
ranks and includes non-related white immigrants. It has given itself a
racial, not a cultural base. After colonialism and fascism were abandoned,
such racial nationalism became unacceptable. If ethnic groups want to
exclude members of their own kind on such spurious grounds, they can
do so only in the private realm. However, Afrikaner racial nationalism
imposes its definition of ethnicity on the rest of the population through
laws. This has made Pretoria an illegitimate government in the eyes of
its opponents, although the regime exercises effective control over its
territory in a recognised sovereign state.

In the ultra-right perception, ethnic and racial nationalism are identical. Conservative Party leader Andries Treurnicht refuses to distinguish between pride of culture and language on the one hand and race consciousness on the other: "This whole cry about racism . . . I've said if loyalty to and love for one's own people and culture and language, if that is racism, then there are lots of racists in this country. Most of us are race conscious, I'm race conscious. I am conscious of being a white man, distinct from the black man and I don't apologise for that."[1]

South African nationalists of all varieties insist on the reality of ethnic diversity in the country. They are fond of calling it a society of many minorities, or a variety of population groups, independent of apartheid. F W de Klerk argues: "There is no way in which all the people of this country will suddenly forget who they are, where they come from, where their roots are and what their aims, motivation and their basic cultural outlook on life is."[2] Such a statement about the historical location of all people everywhere is obviously correct. What it ignores, however, is the other reality of engineered group divisions in South Africa.

Apartheid deliberately arrested trends towards a more integrated society. It reinforced group consciousness where it otherwise would have waned. Above all, apartheid has institutionalised racial nationalism and attached privilege to it at the expense of the disenfranchised. Describing this unequal historical outcome as natural ethnic plurality in which people merely are loyal to their own kind, confuses justifiable ethnic awareness with unjustifiable inequality and enforced discrimination. The "separate but equal" excuse has long been exposed as a myth. In the 1950s the United States Supreme Court rejected the notion as contradictory because separateness is inevitably experienced as rejection and inequality.

Analysts of ethno-nationalism have stressed the kinship symbolism as the secret emotional appeal of ethno-nationalism.[3] Pierre van den Berghe reduces ethnicity to nepotism, an evolutionary conditioning to support kin over non-kin.[4] If indeed the "emotional depth of ethno-national identity"[5] lies in kinship imagery, then the racial nationalism of Afrikanerdom faces an additional problem. By denying brown Afrikaners group membership but allowing white immigrants from around the world to claim that status, Afrikaner nationalism does not recognise kinship. Instead of common descent, it bases the white "community" on a putative racial kinship. That makes the South African dominant group different from communal groupings elsewhere.

South African whites are not an extended family who share ethno-national descent. Instead they form a doubly divided family: divorced from their poorer and excluded coloured real kin and divided culturally between Afrikaners and English. This objective reality of an artificial "family" makes its subjective perception weak and vulnerable. A "white nation" is difficult to sell. Afrikaner nationalists are fond of debating whether the *volk* (Afrikaners) or the *nasie* (whites) should form the ingroup. The Afrikaner National Party has now almost as much support in the English-speaking group as among Afrikaners. The ultra-right Afrikaner Conservative Party is supported by about 12 per cent of English-speaking whites, who comprise about 40 per cent of the white population.

Clearly, however, it is not facts or historical accuracy but people's beliefs that influence behaviour. Therefore, it is an empirical question whether the settler nationalism of the South African ruling group shows a similar emotional bond as ethno-nationalism elsewhere. Such is clearly the case with the minority ultra-right which has now appropriated the discourse and symbolism of Afrikaner nationalism.

The ruling National Party no longer bases its justification for racial group domination on ethnic ideology but on technocratic survival and privilege. The 1989 "plan of action" of the NP never even mentions such loaded sacred terms as race, nationhood, destiny, heritage, ethnicity, Afrikaans, English, African or Calvinism. There is only one brief reference to the "Christian faith" in support of the advocacy of supposedly "common values" shared by 80 per cent of the population. The current identity of groups itself is questioned: "The present basis in terms of which groups are defined for the purpose of political participation creates many problems". The party advocates "freedom of association", meaning that "a person must be able to change to another group subject to the consent of the recipient group". A group for which South African citizenship is the only qualification is also proposed. Since the programme constitutes an election manifesto, the party must obviously expect its voters to share these sentiments.

The interesting question is: does the absence of ethnic legitimation of groups reflect supreme self-confidence in the persistence of racial categories that can be taken for granted? Or does the technocratic reference to "democracy", "civilized norms", "negotiations", "free enterprise", "safety and harmony", and "depoliticised deadlock-breaking mechanisms" in a state "in which cultures and interests differ" herald the search for an overriding new legitimacy?

If the ruling group itself is no longer willing to justify its domination in ethno-cultural terms, does this reflect a shift to an essentially economic class conflict with patriotism and common citizenship as state ideology rather than a nationalist/racial struggle? Can the

National Party transform itself into a non-racial political grouping that would appeal to conservative, law-and-order oriented persons other than whites, thereby expanding its base rather than being confined to a permanent minority? Or will the party merely issue appeals for votes to the newly enfranchised (like it does with white immigrants) but not include them in any significant way?

Patriotism in multi-ethnic states

Patriotism must be distinguished from nationalism. Patriotism is the unifying concept in immigrant societies such as the United States or Canada, the multi-nation state of the Soviet Union or the artificial states of colonial creation. With a variety of groups of different religions and languages, the myth of common origin obviously cannot be invoked.

In such multi-ethnic societies, the creation of the state is celebrated in the flag, oath of allegiance and national anthems very much like in nation states. However, the loyalty demanded from the patriots is not based on a common history but on the unique opportunities that the new "fatherland" provides. The emphasis is on citizenship as a common bond providing equal rewards.

Such promises and demands on newcomers form as cohesive a state as nationalist mobilisation. New immigrants disproportionately volunteered for the US army to show their identification. Shaw and Wong have put the distinction between nationalism and patriotism well. Patriotism, they argue, "is an ideology that promotes loyalty to a society that is territorially and politically defined regardless of the cultural background of its members".[6] Patriotism is therefore more inclusive.

Anthony Smith, without using the term patriotism, has written about a civic ideology of the state which means the same thing. "The civic vision treats nations as units of population which inhabit a demarcated territory, possess a common economy with mobility in a single territory-wide occupational and production system, common laws with identical legal rights and duties for everyone, and a public, mass education system with a single civic ideology." Smith attempts to reconcile the nationalistic/group-based and territorial conceptions of a "nation". He proposes the introduction of "ethnic foundations" to such polities. Ethnic foundations consist mainly of the myth of common origin.[7]

However, genealogy cannot be manufactured arbitrarily, particularly in racially divided societies. There must be some historical foundation to the myth of common descent. Poly-ethnic societies, particularly when they originated from conquest, cannot rely on ethnic nationalism for cohesion, because in those states historical memories are divisive. Traditions are not shared.

If the intellectuals, in the mode of 19th century European nationalists, were to return to ancestral ways and ideals in order to "reawaken the nation" they would promote fission and not fusion. Therefore, poly-ethnic societies can never be nation-states in the strict sense of the term but only modern civic states. These multi-national states tolerate nations in their territory by either granting them some limited constitutional autonomy or by relegating them to the cultural sphere.

If nations "awaken", there is the permanent threat of separatism, that is, the end of a unified state. It is this prospect that haunts the opposition in South Africa. Recognition of ethnic group rights would lay the seeds of separatism and irredentism, as indeed the programme of the ultra-right does. At the most, a policy of symbolic multi-culturalism such as propagated in Canada and Australia can reconcile heritage maintenance with patriotic loyalty to the multi-ethnic state.

However, the ruling group has now in principle conceded the civic vision of patriotism. The crucial turning point was the acceptance of Africans as permanent citizens in the urban areas in the late 1970s. That meant the end of the dream of partition through grand apartheid. The National Party, however, has so far failed to concede the full implications of common citizenship. By insisting on the untenable distinction between "own affairs" and "joint affairs", Pretoria tried to keep racial control of the central government.

This political schizophrenia lies at the heart of South Africa's legitimation crisis, as the separatist conservatives and non-racial democrats have both stressed. This twilight status of a patriotic state in the making has also prevented the emergence of unifying symbols and shared feelings of belonging to a common South Africa. If such a new non-racial "nation" is to develop, it cannot rely on ethnic histories but must look to a common future for its cohesive glue. Since the past shapes the future, the project of South African patriotism faces unprecedented challenges.

Many analysts have rightly noted that "in the political and cultural sense there is no nation in South Africa".[8] The society is "peculiarly devoid of national symbols"[9] that would have meaning for all its inhabitants. This state of affairs obviously results from successive governments since 1910 elevating their sectarian claims to national status. The absence of one South African nation, however, does not necessarily reflect allegiance to sub-nations on the part of the majority, as is the case in other divided societies. The South African majority would want to create one non-racial "nation", if it could only break the monopoly of political power of the minority. An inclusive patriotism could obviously fill this ideological vacuum and provide the cohesive glue.

Patriotism, too, has been falsely abrogated by Afrikaners as loyalty to the *volk* only. The ultra-right named its defunct mouthpiece *Die Patriot*. Equally misleading can be the elevation of patriotism to the tyranny of "love it or leave it". Flag-waving patriots can be as dangerous as militant nationalists. In this sense, patriotism represents mere territorial nationalism.

However, true patriotism tempers loyalty for the fatherland/motherland with an awareness of internationalism and global interdependence. It seeks unity in diversity. Individuals and groups from different backgrounds and with conflicting interests develop common values and political rules despite differences. Just as South Africans have already achieved what may be called "industrial patriotism" — commitment to negotiations and rule-bound bargaining between class antagonists on the basis of free association and equality — so patriotism in the political sphere could promise stability and new opportunities. This would not exclude variety in cultural practices, official multilingualism for example, but would override different traditions by means of common rights rather than institutionalising ethnic differences.

These problems do not exist in Israel, which has rejected incorporation of the territories precisely because it would mean extension of citizenship to their inhabitants. Israel has chosen to remain an ethnic state. Its nationalism (Zionism) cannot be reconciled with an ideology of patriotism as far as the Israeli Arabs are concerned, although they enjoy nominal citizenship because they constitute only a 15 per cent minority. On the other hand, decolonising the West Bank and Gaza and allowing an independent Palestinian state would mean not only abandoning the dream of territorial Zionists for a Greater Israel, but also providing a focus for the emotional attachment of Israeli Arabs.[10] Separation of two competing nationalisms seems the only realistic answer here, while partition remains unfeasible in South Africa.

It is the central thesis of this analysis that the African National Congress (ANC) promotes patriotism. The South African issue is therefore not a conflict between Afrikaner and African nationalism, rather the ANC offers patriotism as an alternative to Afrikaner nationalism. The strife in South Africa is therefore not comparable to nationalist conflicts in other divided societies, such as Israel and Northern Ireland. In terms of this reasoning it is justifiable to be much more optimistic about the outcome of the South African strife than the more intractable nationalist conflicts elsewhere.

The fallacy of bi-communalism

Analysts often falsely apply lessons from intractable disputes elsewhere to South Africa. They all start from the assumption of a communal conflict in which historic communities wish to preserve their unique identity. Bernard Crick, for example, states unequivocally that "individuals in South Africa find their human identity in being a member of a specific group" (see Chapter 15).

Such a statement is either true universally, in so far as everyone is a member of his/her group, and therefore platitudinous, or it applies to South African racial groups, which makes it questionable. Indeed, it amounts to projection of supposed identities on to the majority with little empirical reality behind it — an operation reminiscent of the National Party. In consequence, proposed resolutions to the communal conflict assume "equal justice" to both sides.

Morally, this approach may be applicable to Northern Ireland and Israel/Palestine but it is clearly untenable for South Africa. Even as a mere political device for reconciling communal identities in a common political framework, the notion of equal justice breaks down when one side rejects its own supposed group identity. Consequently, Crick's conclusions are questionable when he asserts: "In South Africa any compromise will be between communities". It would be more appropriate to say: any compromise will be between racial nationalists and the majority committed to individual patriotism, regardless of community affiliation. For this group, identity will be a private affair, while the public realm will be governed by recognition of equal individual rights.

The focus on communal conflict ignores the fact that not even the traditionalist Inkatha movement advocates constitutional racial or ethnic group guarantees. When in a divided society the overwhelming majority of the people favour a non-racial, non-ethnic constitution based on individual rights, the conflict is better conceptualised as the hold-out of a minority against the socio-economic implications of majority rule, rather than as a communal conflict with equal rights on both sides.

What needs to be addressed is not the institutionalisation of ethnicity in a post-apartheid state but how the nationalist minority of ultra-rightists can be neutralised and prevented from wrecking democratic rule. On this crucial question, the ANC has shown itself as un-imaginative, evasive and self-deceptive as the liberal opposition. Bi-communalism at least suggests a potential interim compromise. However, such an arrangement neither satisfies the essential demands of the ultra-right (total sovereignty) nor the core of the opposition goal (non-racialism).

Adherents of the communal conflict conceptualisation argue for parity of representation on executive and legislative levels between a white and a non-racial grouping. This bi-communalism, at least during a transitional period, is propagated in order "to build a new nation over time which would transcend both Afrikaner and African nationalism".[11]

But far from transcending nationalism, constitutional racial dualism would institutionalise and perpetuate invidious group affiliations. Admittedly it would be an advance on the present situation where non-racial or "groupless" representation does not exist at all. A transitional period might also be useful and necessary as an educational device. However, the disadvantage of further uncertainty and instability with another constitution of doubtful legitimacy would clearly out-weigh the advantages of luring more whites into political concessions to the non-racial cause. Racial bi-communalism seems as illegitimate as a qualified franchise: full of good intentions but not realisable in practice.

Even if it were true, as Giliomee asserts, that "virtually all political parties, trade unions and other public associations are communally based and emphasise the promotion of communal interests",[12] this is not due to free choice but out of necessity as a result of legislation. Advocates of communal conflict theory therefore accept an imposed reality as the basis of their solution. Their arbitrary assumptions rule out alternative compromises which start from different definitions of the problem. In short, the folly of the conceptualisation of strife in South Africa as a communal conflict lies in adopting a minority definition although the majority clearly reject the apartheid notion of competing nations.

The South African state may not be challenged by the subordinate population solely in terms of their exploitation, but it certainly is not primarily challenged in terms of an alternative exclusivist nationalist movement. On the contrary, it is by means of an inclusive patriotism that most of the subordinates in the Congress tradition attempt to dismantle their exploitation.

The inhabitants of the townships and squatter camps could not care less about "culture", "national identity" or "heritage maintenance", if they could only get jobs, decent housing and equal pay through political equality. Their protest against "national subjugation" — the denial of equal rights, in other words — is only a means to end discrimination, not an end in itself. That does not mean, however, that the majority can be bought off with material improvements, as the co-optation policy attempts. Economic exploitation and national oppression are inextricably linked. The attempted capture of state power aims at rectifying exploitation.

Unlike Isarel/Palestine and Northern Ireland, where two mutually exclusive nationalisms confront each other, in South Africa racial nationalism is challenged by liberalism and a variety of "socialist" visions. This camp comprises adherents of various cultural traditions. Even the ruling Afrikaner nationalists are increasingly forced to co-opt members of other racial groups into their formerly exclusive realm. This further diffuses a racial conflict into a dispute between privileged powerholders and excluded subordinates in the same state.

Due to the heterogeneous composition of each camp, intra-group conflicts frequently override the intergroup divide. The intensity of strategic disputes between the government and its ultra-right opposition testifies to this constellation as does the fighting between the anti-apartheid Inkatha and the ANC/United Democratic Front and African-ist tendency. The concept of a "white community"[13] is therefore as questionable as that of a black community. It makes a caricature of the deep political cleavages between Afrikaners to attribute to them "a vivid sense of oneness of kind".[14] The advocates of "national dualism" grossly neglect these cleavages in South Africa.

The alleged need for "a set of dual national symbols" is therefore empirically questionable.[15] It is doubtful whether post-apartheid society will produce "a liberated nation of and for Africans" where "everyone will have to defer to African symbols". Obviously, the present Afrikaner monopoly of symbolism will have to be supplemented by African values. "Die Stem" will give way to "Nkosi Sikelel' iAfrica", Jan Smuts Airport will become Mandela Airport and there will be many June 16 streets across the land.

However, the Voortrekker monument is unlikely to be blown up, Dutch Reformed churches will still be full and education in Afrikaans will continue, although the biased history curriculum will have to be rewritten. The ethnic character of most small towns and neighbour-hoods will hardly change, even after the Group Areas Act has been repealed. The authors who worry that "the numerical preponderance of blacks will unlock the symbolic ideal of a liberated African state"[16] paradoxically also admit "the fact that almost all white fears of majority rule are based on myth".

While one should not necessarily accept the ideal of non-racialism at face value, one should also not dismiss the many ANC statements of support for a "common loyalty and patriotism" as mere propaganda. In Albie Sachs's attractive vision the goal is not to promote "identikit citizens". The objective is "not to create a model culture into which everyone has to assimilate, but to acknowledge and take pride in the cultural variety of our people".[17] As long as such a non-racial multi-culturalism is not linked to privilege in the public arena, it promises to reconcile the different collective histories rather than to deny them.

Assertions about a common society in South Africa, however, need to be further substantiated with a closer analysis of the current state of the white and black mind. What is the essence of the strategies and concerns in both camps and how far do the main political organisations historically and presently reflect nationalist visions or the logic of material interests?

Afrikaner nationalism reconsidered

The colonial residues of settler conquest obviously still exercise a powerful influence on how whites (and blacks) think about themselves and structure their relations. But the ruling settler descendants are also now increasingly concerned with securing their material advantages from this history, rather than preserving their collective identity. Legal racial stratification, in the government's own words, has become an "albatross around its neck" because it obviously now endangers prosperity and political stability. When 70 per cent of the group are considered to be middle class, avoiding jeopardising this position vies with ideological relics.

Ethno-nationalism usually wins when it competes with class mobilisation. However, this empirical observation in the ethnic relations literature always derives from a case of a threatened working class or a downwardly mobile petty bourgeoisie that compensates for denied aspirations with symbolic status. It has yet to be proven anywhere that a BMW-owning bureaucratic bourgeoisie with swimming-pools and servants readily sacrifices the good life for psychologically gratifying ethnic affinities. Racial sovereignty proves durable only so long as it can deliver. A bureaucratic oligarchy can be expected to drop its "albatross" when racialism becomes dysfunctional.

In truly nationalist conflicts, a group identifies with the homeland or state as an exclusive property. But in South Africa the ruling minority shares the territory with a vast majority, and control of the state is not so much a matter of identity for most whites as a source of blatant self-enrichment. The NP government regularly buys its vote through pre-election handouts. A dramatic increase in corruption scandals among government officials also testifies to an instrumental rather than emotional use of the state. It is this access to spoils and a privileged lifestyle more than identity which the South African regime does not want to lose. If it needs to co-opt auxiliaries from other racial groups or even share some of the spoils, it is ready to oblige.

Expediency at the top affects the bottom as well. Increased opposition to army service and emigration by professionals is only one indication of weakening commitment to nationalist sacrifice in South Africa. In Israel, on the other hand, 90 per cent of Israeli youth would volunteer were they not drafted, as a recent survey for the Ministry of Education revealed.

Emotional identification with an ethnic cause can also be gauged from the position of its writers and intellectuals. Poets, artists and singers have always been the prime articulators and mobilisers of nationalism. Despite vast political differences among Israeli writers on the liberal/left spectrum, few disagree with the Zionist core philosophy of maintaining Israel as a Jewish state. Regardless of the government of the day, there is unwavering endorsement of this Jewish nationalism by the Jewish population as well as by most of its intellectuals.

In South Africa, on the other hand, many of the Afrikaner intelligentsia have defected from the nationalist cause. Many Afrikaans writers and poets now sing the praises of the ANC. They identify with a common patriotism. What is left of academic support for the old order is of the social engineering kind. In the words of the cynical leader of this group, Piet Cillie, apartheid had to be tried in order to prove that it could not work.

The NP managers display smug bemusement at the quaint nationalist antics of the ultra-right. Opposition parties on both the right and the left criticise the government mainly for economic mismanagement and no longer for ideological betrayal. They bank on a protest vote, not on the superiority of their own vague programmes.

This strategy reflects the clear priority for the electorate of economic issues. Concerns about identity rank far below worries about inflation, pension and currency values.[18] The conventional wisdom that whites act out of justifiable fear of a potential black takeover needs to be revised, therefore. The fear is better described as anxiety about losing the lifestyle to which they are accustomed.

Closely related is an anxiety about crime. In the white mind, losing political control means also being deprived of physical protection for property — a responsive police force of their own. John Coetzee has suggested that the white fear of theft of their property emanates from the subconscious knowledge that what they have is itself stolen, that they have no real right to it.[19]

One does not have to view the South African conflict as a disguised class struggle to notice its essential economic nature. A privileged multiracial minority under the political leadership of an Afrikaner bourgeoisie and state bureaucracy, with the tacit connivance of ideologically fragmented English capital, faces a vast proletarian demand for a better quality of life through political enfranchisement.

However, the struggle between a developed white sector and a grossly neglected black majority is distorted by the alignment of three major class actors with the opposite side. Firstly, some big business interests and affiliated middle-class professionals support the dismantling of racial domination and the institution of social democracy (but not socialist restructuring). Secondly, a declining white working class has made common cause, under ethnic auspices, with the more extremist elements of the apartheid state. Finally, newly urbanising unemployed migrants in huge squatter camps also back status quo forces under the coercion and patronage of local warlords. The forces of the status quo are augmented by a growing number of blacks in the civil service and vast apolitical segments with strong law and order leanings.

In its Constitutional Guidelines, the ANC commits itself to a "rapid and irreversible redistribution of wealth", but stops short of specifying how this worthy goal can be achieved without major economic disruption. Obviously major redistribution can be envisaged only in the context of economic growth. Ignoring independently operating market forces hostile to well-intentioned state intervention could reduce redistribution to spreading poverty more equally.

An economy is far more difficult to manipulate than a constitution. The highly capital-intensive South African economy remains in a crisis that is structural, not merely cyclical, due to its dependency on expensive imported machinery, an underdeveloped domestic market and skewed racial social costs. While a political domestic settlement would obviously greatly improve economic prospects by freeing hitherto restricted resources in a new psychological climate of optimism, the delicate trade-off between growth and redistribution would remain a problem in a post-apartheid economy. What matters are not slogans like "mixed economy" (which all economies are), but the quality of the mix and its differential benefits.

Naturally, whites are likely to have to make material adjustments in a post-apartheid society. There will be higher taxes in the interests of equalising welfare payments. There will be affirmative action programmes, particularly with regard to the civil service. If an alternative government were to shy away from such basic redistribution, it would betray its main constituency.

However, the economic boom which certainly will follow a political solution is also likely to offset much of the equalisation costs in the long run. Continuous apartheid, on the other hand, would escalate poverty in a strife-ridden siege economy, further emigration and deepen the isolation of an increasingly backward country sliding into irreversible social and economic decline.

This basic line-up is fluid and can change in accordance with a multiplicity of events and policy initiatives. Competition for electoral support on its left, Eastern European developments as well as outside pressure has induced Pretoria to be more susceptible to a negotiated settlement. Likewise, the retreat from socialist experiments world-wide and superpower collaboration on regional conflicts has forced the South African opposition to adopt a more accommodating attitude to eventual negotiations.

Postscript

The liberal euphoria about the normalisation of politics after 2 February 1990 and the release of Mandela overlooks the fragile basis of this momentous change of course. Similarly, the black perception that whites en masse have abandoned racial identity and happily embraced the new South Africa deceives itself.

For the first time in South African history the government finds itself far ahead of its constituency. It leap-frogged decades of indoctrination. The white electorate as a whole has not yet realised the implications of normal politics. Both the NP and the ANC have failed to educate their following in the art of compromise. When state-controlled SATV suddenly gives lengthy and almost sympathetic exposure to ANC spokespersons and exiles arriving with clenched fists at South African airports, the result is massive confusion. Anxiety is reinforced by the flag of the South African Communist Party flying at ANC press conferences and mass rallies, when in Eastern Europe similar crowds cut out the hammer and sickle as symbols of oppression.

Increased gun sales and a plummeting stock market indicated a political ignorance that took literally Mandela's stated support for nationalisation and a symbolic armed struggle. Isolated incidents of looting by brutalised township youth, widespread upheavals in homelands and a tarnished security establishment confirmed the latent unease.

Frequent reassurances by Mandela and the ANC about the role of private business and political inclusion ("structural guarantees" in Mandela's first speech), contributed little to allay apprehension because a substantial minority does not believe that Mandela can control the townships.

Results from a snap opinion poll conducted by *Rapport* during the week after Mandela's release confirmed the split reaction among whites. Instead of renewed hope and relief, optimism decreased from a 50 per cent figure in November 1989 to 40 per cent. Afrikaners in particular showed more pessimism about the future, with a 59 per cent "optimistic" response dropping to 31 per cent three months later. Optimism among English-speakers by contrast increased from 38 per cent to 53 per cent. Increased support for the ruling National Party (from 41,6 per cent to 46,3 per cent) came at the expense of the liberal Democratic Party (18,3 against an earlier 24,3), while support for the Conservative Party increased from 22,6 to 31 per cent. The white backlash notwithstanding, negotiations and compromise are on course. For the first time Pretoria is heading in the right direction rather than into a cul de sac.

The interesting question remains whether official power-holders can determine the outcome of their new approach. They themselves display little vision of where their *ad hoc* management should lead. Just as Gorbachev's perestroika developed its own dynamic of releasing ethno-nationalism in a disintegrating Soviet empire, so Pretoria's new approach may simultaneously strengthen as well as weaken the opposition. In the new interplay between previously implacably hostile sides both antagonists change and emerge with new strategies.

Developments in Eastern Europe obviously influenced strategic shifts on both sides in the South African conflict. The ANC has been deprived of its previous support base. The collapse of Stalinism together with the discredited Leninist notion of a vanguard party finally prompted even the South African Communist Party to opt for democratic socialism with a multi-party system and formerly despised liberal freedoms. The self-confident regime therefore no longer perceives the opposition as a major foreign threat that would impose a Marxist dictatorship of the proletariat with Soviet assistance. The white mobilising notion of a "total onslaught" became discredited with the end of the Cold War. New competition with Eastern Europe for Western capital provided the final trigger for a political solution as a precondition for attracting foreign investors.

Moreover, the revival of ethno-nationalism in the Soviet Union is considered proof of the power of ethnicity, even by progressive academics. However, South African ethnicity differs fundamentally from its alleged counterpart in Eastern Europe. Under Stalin ethno-nationalism was suppressed. Under apartheid, ethnicity was promoted by the state to fragment the majority. This state policy has immunised most of the politically conscious against a revival of tribalism. Above all, Pretoria fostered a racial not a cultural nationalism for whites. Ethnicity became the reactionary mark of inclusion or exclusion.

In the Soviet Union on the other hand, ethnic mobilisation to a large extent is associated with liberation from Russian domination. Progressive regional nationalism in the Baltic States cannot be compared with a reactionary defence of racial privilege in South Africa. Because of its moral base there and illegitimacy here, ethno-nationalism can indeed be expected to grow in the Soviet Union, but to decline in the apartheid state. In short, ethnicity again proves to be circumstantial and situational rather than primordial.

The danger for South Africa lies in the temptation to create the semblance of a formal democracy, with a universal franchise and veto powers for entrenched interests that leave the enormous economic inequality basically intact. As long as political freedom does not translate into substantial economic restructuring, the appeal for socialism remains strong in South Africa, Eastern Europe notwithstanding. Without addressing material inequality, South Africa is likely to decline into Brazilian conditions with crimes of poverty invading the privileged enclaves or permanent repression as a response.

Unlike ethnic conflicts elsewhere, the South African conflict is fortunately not perceived in religious and value terms. Essentially a struggle over racially allocated power and privilege, the contest remains open to bargaining. Northern Ireland, Lebanon, Israel or Sri Lanka remain far more intractable because differentially allocated life chances are tied up with religious identity.

ENDNOTES

1. *Leadership*, Vol 7, No 6, 1988, pp8-11.
2. *SA Report*, 15 June 1989.

3. Walker Connor: "Ethnonationalism", in Weiner and SP Huntington (eds): *Understanding Political Development*, Boston, Little, Brown and Co, pp196-220. Donald Horowitz: *Ethnic Groups in Conflict*, Berkeley, University of California Press, 1985.
4. Pierre van den Berghe: *The Ethnic Phenomenon*, New York, Elsevier, 1983.
5. Connor: "Ethnonationalism", p204.
6. R Paul Shaw and Yuwa Wong: *Genetic Seeds of Warfare: Evolution, Nationalism and Patriotism*, Boston, Unwin Hyman, 1989, p158.
7. Anthony D Smith: "The Myth of the 'Modern Nation' and the Myths of Nations", *Ethnic and Racial Studies* Vol 11, No 1, 1988, pp1-26.
8. Neville Alexander: "The Language Question", University of Cape Town, Institute for the Study of Public Policy: Critical Choices for South African Society, No 12.
9. Peter Berger and Bobby Godsell: "South Africa in Comparative Context", in Berger and Godsell (eds): *A Future South Africa*, Cape Town, Human and Rousseau, 1988, p281.
10. Mark Heller: *A Palestinian State*, Cambridge, Harvard University Press, 1983.
11. Hermann Giliomee: "The Communal Nature of the African Conflict", in H Giliomee and L Schlemmer (eds): *Negotiating South Africa's Future*, Johannesburg, Southern Book Publishers, 1989, p127.
12. Giliomee: "The Communal Nature of the SA Conflict", p114.
13. Hermann Giliomee and Lawrence Schlemmer: *From Apartheid to Nation-building*, Cape Town, Oxford University Press, 1989, p226.
14. Giliomee: "The Communal Nature of the SA Conflict", p118.
15. Giliomee and Schlemmer: *From Apartheid to Nation-building*, p239.
16. Giliomee and Schlemmer: *From Apartheid to Nation-building*, p240.
17. *Cape Times*, 14 February 1990.
18. Lawrence Schlemmer, in *Sunday Times*, 13 September 1989.
19. Personal communication, letter of 12 September 1989.

14 Strategies for the future
LAWRENCE SCHLEMMER

The "end of history?"

The perceived decay of socialism as a viable model for economic development and modernisation, and more recently the swift collapse of communist regimes in most of Eastern Europe, have induced widespread anticipation of a new world era of reconciliation between East and West, and the eclipse of major ideological tensions. Francis Fukuyama has responded to these shifts with his much debated and challenging proposition, "The End of History?".[1]

The notion that the "grip of history" will be unlocked, and the possibility of a benign, perhaps even boring (according to Fukuyama) international agreement over public values and economic methods has parallels in the current phase in South African politics.

In recent years a situation has emerged in South Africa suggestive of a morally defeated government in retreat, attempting to repair its failures, and of the African National Congress (ANC) waiting in the wings, supported by a surgent human rights movement, steadily advancing to ever higher strategic and moral ground — a classic model of political conflict between new and old, good and bad. The publicity successes of the ANC in this situation probably exceeded even its own expectations.

The product of a strategic alliance in the fifties between an African nationalist movement and white, coloured and Indian progressives and socialists, the ANC had the additional advantage of a basis for mobilisation — progressive non-racialism — which had enormous appeal to intelligentsia, journalists and academics throughout the West. In the context of international liberation movements with distinctly uncomfortable images — Muslim fundamentalism, other kinds of religious sectarianism, Latin American revolutionary ideology, anti-

Zionism — the ANC's apparently benign and ambiguous socialism of the Freedom Charter and its universalistic non-racialism have been like symbolic manna from heaven for idealistic intellectuals and professionals in South Africa and the West. All kinds of lobbies and voluntary change organisations have become added to the older anti-apartheid and socialist movements in the orbit of the ANC's progressive non-racialism.

A powerful, intellectually contagious mood of expectation, akin to the mood fired initially by Gorbachev's glasnost and the breaking of the Berlin Wall, has burgeoned in the wake of more and more political concessions by the South African government.

Increasingly the positive expectation of a non-racial democracy in South Africa infuses the communication of not only academics and the liberal white opposition party, the Democratic Party, but keynote white industrialists, the liberal media and a wide range of ordinary white moderate reformers as well.

The recent actions of President F W de Klerk — releasing Nelson Mandela, unbanning the ANC, tolerating most protest marches, opening the door to negotiations with the ANC and even using the words "new South Africa" himself — have shifted the climate of debate from expectations to near-certainties. Cautious "realpolitik" analysis is often seen as irrelevant or retrograde. The remaining race-based statutes are seen as infuriating residues certain to be washed away in the months that lie ahead. The political tendons in the "grip of history" are seen to be snapping, and its institutions and values crumbling.

We must note, however, that Fukuyama has no particular expectations of the passing of sectional passions in small states on the margins of the great international consensus. "Palestinians and Kurds, Sikhs and Tamils, Irish Catholics and Walloons, Armenians and Azeris will continue to have their unresolved grievances". The question is whether South Africa will join the world trend toward Fukuyama's "homogenisation", or whether the divisive dynamic of its history will endure. Is South Africa's history ending?

If one takes a very long view, and if one is indefinite about time scales, one may fairly confidently proclaim that the process of national unification is irrevocable and the end result, majority control of government, inevitable. The groundswell of internal majority expectations, the strain in public institutions forged on minority rule, foreign pressures and more recent demonstration effects from Eastern Europe have all helped to fragment the formerly rock-like political edifice of "European" hegemony which stretched back for more than three centuries.

In the short and medium term, however, these optimistic expectations may be facile. It is conceivable that historical conflicts will re-emerge, perhaps particularly if the residues of apartheid are merely swept under some carpet of hopeful expectations about what a modern and rational country should be like.

Alternative futures

South Africa's political pathways into the future are by no means sign-posted yet. The following are theoretical alternatives which have to be held in analytical consciousness at this stage.

Negotiated transformation into majority rule

The essence of this future is built round the expectation that powerful majority sentiments, democratic world opinion, the determination of the ANC, declining commitment to sectional political interests in the National Party and their replacement by an interest in economic growth unfettered by sanctions and political stress, will all lead to negotiated majority rule.

Certainly most realistic observers would expect some protections and guarantees for minorities, but the principle of undifferentiated democratic participation will be the keynote. It is this view of the emerging future which is shaped by the rising perception that a non-racial ethic will become dominant; that any alternative to this is at best archaic and at worst perverse. It is a feeling about the future which in recent months has permeated boardrooms, white professional circles and the media.

It is comforting and reassuring, and many categories of political and economic fears among whites are assumed to be able to be reduced by the goodwill which will characterise the new society. Consequently, conservative resistance is not assumed to be capable of challenging the transformation. For black South Africans this is understandably a fervent ideal.

Negotiated compromise

This is an anticipated resolution of a somewhat more "realpolitik" kind. Broadly it relates to the simple assumption that both the minority and majority principles are likely to be in balance for some time.

It assumes that the majority principle has great popular credibility and is a condition for the success any resolution will achieve in ending

246 • LAWRENCE SCHLEMMER

political conflict. However, it also takes account of the fact that the present government, as the dominant voice of white minority interests, has all the leverage of an established administration and, both in terms of its own commitments and because of competition from its right, will insist on substantial minority participation in government or a sharing of executive power between majority and minority formations.

This outcome can be seen as either a longer-range future or an arrangement which will be some form of interim bridge to a future resembling the majority rule scenario.

A return to political separation

South Africa's transition will be difficult. As one sees already, mass expectations are high and the ANC and other majority-based movements, understandably, are likely to organise or at least utilise political instability to improve their bargaining position in negotiations. The government will run the risk of a breakdown of its image of control over an orderly transition if mass protests escalate. Yet at the same time its security action will be constrained by attempts to meet the ANC's conditions for negotiation.

The emerging situation will place great pressure on the government's cool-headedness and crisis management. If it appears to be losing its grip on the situation, as already speculated on by columnist Simon Barber (*Business Day,* 10 April 1990), or if white reactions and protests in response to uncertainties mount, the government may be forced to call an election at an inopportune time before 1994, the normal date for the next parliamentary election.

The right-wing opposition party currently has substantial white support and under conditions of uncertainty could augment its support sufficiently to narrowly win an election. A right-wing government would undoubtedly use massive repression to bring unrest under control and would reinstitute a policy of political and geographic segregation, aimed at least at preserving a major part of the country for exclusive white control (albeit smaller than currently, perhaps).

Amid mounting international sanctions, a barely contained revolutionary climate and a sluggish or faltering economy, the country would revert to a new attempt to make apartheid work. This scenario would not endure in perpetuity but it could extend into the medium-term future.

White capitulation

Under conditions such as those described in the previous scenario, the government might lose its nerve and attempt to achieve stability and to encourage the ANC to negotiate by making sweeping concessions, as has been the tendency of the communist parties in a number of East European countries recently.

Characteristically, however, such concessions heighten expectations and increase pressures. Hence a logical outcome would be a hasty formula for withdrawal and a rapid installation of a majority-based government under conditions of social hostility and economic breakdown.

An expected outcome would be a radicalised majority government which would damage established economic institutions in its haste to achieve socio-economic redress.

This scenario includes the assumption that whites will become so demoralised that a core of white resistance, with support from security agencies, will not materialise; that the present South African government is the last bulwark of white minority interests. These are assumptions which must be assessed carefully.

Stalemate

There is another scenario based on an untidy mix of the elements above which is simply an extended stalemate, in which negotiation is protracted and no concessions are made. Minority and majority interests are held in balance in a situation of increasing social ferment.

These alternative short- to medium-term futures are not equally probable, but are all possible. They are offered as a framework within which to examine strategic requirements in South Africa's transition; requirements which are often obscured by topical analyses based on mood and media communication.

Whichever future emerges in the medium term depends on two broad sets of factors. The first set almost defies systematic analysis. It is the political poker game of negotiations, composed of the strategic moves and countermoves made by the major protagonists in the brittle process of transition, their capacities to mobilise pressure and resources, manage crisis, form pacts, the degree of emerging trust and their capacities to both control, motivate and maintain coherence among followers. Assessing this level of social action is essentially short-term, and while fascinating and obviously important, demands immediate and ongoing evaluation and comment, most appropriate in the daily media.

The second set of factors are the more fundamental and underlying issues which have a more consistent and continuous effect on the outcomes, either by constraining options or creating opportunities. These factors are made up of institutional patterns, values, interests and socio-political and cultural needs. Less unpredictable than the first set, these underlying dynamics have effects which can be anticipated and are less topical, and hence amenable to systematic study.

These latter factors are the "grip of history" and it is these elements which will form the basis of the analysis which follows. Given the tragedy which some scenarios imply, a correct analysis is vital. Of these factors one is most salient: majority versus minority interests in a negotiated resolution, or in other words, how to dismantle apartheid.

Majority versus minority interests

Formally established and legalised political and social discrimination between groups based on race, taking the form of the exclusion of the majority from meaningful participation in mainstream political institutions and spheres of social life is what all South Africans and the world know as apartheid. The political conflict in South Africa is dominantly about apartheid and it is utterly essential that apartheid should be destroyed. Any attempt to retain this kind of system will undoubtedly deepen conflict and create the danger that a scenario of violence and economic destruction will follow.

These obvious points notwithstanding, the question arises as to whether or not some form of accommodation of minorities, at least in an interim dispensation and possibly in a negotiated constitution, is appropriate. This is the position of the National Party as it enters negotiation. This position has not been fully explicated but it is fairly evident that it is likely to differ from the present constitution, with its distinction between "own" and "general" affairs, and the guaranteed dominance of the major white party in Parliament.

The government's negotiating agenda is being kept as open and unstructured as possible, but all comments and signals point to the government's negotiating position being some form of established coalition between itself and the majority party, probably secured by one or another form of over-representation of minorities in Parliament. This position is unlikely to be accompanied by any demands for formal social discrimination in the current sense. Recent government moves to allow "local option" in schools and residential areas, the existence of a committee investigating the Land Act and indications obtained in personal conversations all suggest that a sustained process of incremental desegregation will occur from now on.

Hence the purpose of minority participation in a new or interim political arrangement will be very largely to provide back-up reassurances for whites who might not otherwise be willing to accept majority rule, to safeguard very basic cultural rights, such as Afrikaans as one of the official languages and the right to "voluntary dissociation" in the form of ethnic schools for those who want them. Over and above these motivations, Constitutional Affairs Minister Gerrit Viljoen has articulated a major consideration: the need for "quality of government", which could be facilitated by built in checks and balances between majority and minority representation. (*Saturday Star*, 17 March 1990).

In other words, the intention behind government thinking probably is to arrive at the second scenario (power-sharing) through a negotiated retention of minority participation. Needless to say, this is not what the ANC/Mass Democratic Movement (MDM) would hope to achieve. Not unexpectedly for a formation representing majority interests, they would wish to secure as much untrammelled power for a majority-based party as possible. The ANC has also been through decades of struggle during which it developed a dual political theory, one aspect informed by communist revolutionary theory, and the other by the tradition of total liberation and complete takeover of power by the resistance movements in colonial Africa. Neither tradition acknowledges any legitimacy for existing power-interests.

In the normal course of events, therefore, one would expect the government and the ANC to enter negotiations with fairly polarised positions, and for the outcome to represent some form of compromise. Such a compromise might be proportional representation of parties without reference to groups in the legislative assembly (majority principle) but with some representation for minority parties on the cabinet and/or over-representation of minority-based parties in a second chamber (consociational principle).

If there is a sustained and determined bid by any formation to rule out the negotiating position of their major opponents, given the very binding commitments of the protagonists to constituencies the result is not likely to be a victory for one or another position, but the collapse of negotiations.

There are prospects of some sort of mobilisation against either the "majority" position or the "consociational" or minority position. The former would be in the form of white grassroots mobilisation; the latter would be attempts by the ANC and its allies to discredit and rule out the consociational position by depicting it as "apartheid".

As regards white resistance, one must bear in mind that F W de Klerk, by committing himself to achieve a compromise with the ANC, is stretching his mandate quite considerably. In a recent nation-wide poll (Report No 1/90, February 1990), Market and Opinion Research found that only 54 per cent of National Party supporters and 44 per cent of all white voters "approved" of the unbanning of the ANC, Pan Africanist Congress (PAC) and SA Communist Party. The same poll showed a 12 per cent increase in support for the Conservative Party (CP) between November 1989 and February 1990. This and other indications suggest that the level of support for the CP in a hypothetical election would be at least 38 per cent of the white vote.

There is minimal support among whites for the principle of a majority-based legislature. There are numerous examples of empirical evidence to show that while popular white attitudes reveal considerable variation in regard to *ways* of coming to terms with black political aspirations, *near consensus* emerges in the rejection of majority rule or an unqualified "one-person-one-vote" system. A majoritarian system, quite correctly, is perceived as making the organised and effective articulation of white minority interests, as a consolidated position, difficult or impossible.

For example, a 1988 Pretoria-Witwatersrand survey among all races and covering 500 whites, conducted by Marketing and Media Research, established that:

> No more than 11 per cent of English-speakers and three per cent of Afrikaans-speaking whites endorsed a "mixed-race parliament with the majority party in control";
>
> The same proportions as above — 11 per cent of English-speakers and three per cent of Afrikaans-speakers — considered that the government of the country should "change immediately to a non-racial system";
>
> 73 per cent of English-speakers and 70 per cent of Afrikaans-speakers favoured the constitutional protection of "group rights".

Note that the sample was probably more "liberal" than a nation-wide sample of whites would have been, due to its metropolitan bias.

Not only National Party supporters but even the "liberal" opposition, the Democratic Party supporters, clearly seem to require protection against becoming outnumbered by a political majority. Table 14.1 shows the results of a May 1989 national survey by Market and Opinion Research (n 1638) with regard to choice of political options by supporters of the different parties.

Table 14.1: Support for political options:

	NP	DP	CP
Unitary state, one person one vote	2%	5%	2%
Negotiated regional federation with a bill of rights	34%	73%	13%
Power-sharing but with group autonomy in regard to own affairs	40%	13%	12%
Partition	11%	1%	66%
Other or none	13%	9%	7%

The National Party response is what one would expect. The Democratic Party supporters are interesting. They would appear to be emphatic in wishing to move away from racially based participation, but at the same time, in opting so overwhelmingly for a regional or geographic federation with an entrenched bill of rights, they appear to seek protection against the eclipsing effect of undifferentiated majoritarianism.

Given this evidence, it cannot be ruled out entirely that at some stage in the future there could be various manifestations of a sufficiently strong white rejection of ANC demands to prevent De Klerk from meeting the ANC position in a midway compromise. Already there is evidence of a mode of confrontational resistance emerging among white groups on the far right (*Rapport*, 8 April 1990). These groups may be marginal at present but they are no less marginal than the militant wing of the Irish Republican Army (IRA) is in Northern Ireland. Militant marginal groups have influence beyond their activist numbers.

An equal and opposite danger exists on the other side. The ANC has a long established position of regarding any party political mobilisation on the basis of race as something to be prohibited by law. There will thus probably be an inclination to want to depict the National Party's negotiating position in the same light, and hence an attempt to discredit it sufficiently to avoid having to reach a compromise with it. As recently as January, an ANC delegation to a meeting in Botswana told a Youth for SA delegation that in their view the National Party, Inkatha and the PAC should be banned on the grounds that they promoted racism or ethnicity (*Star*, 1 January 1990).

Obviously the ANC has already relinquished this position by agreeing in principle to negotiate with the National Party and by almost having a meeting with Chief Buthelezi (called off due to violence in Natal). This, however, does not mean that groupings within the broad

ANC/MDM alliance will not mobilise against any compromises with the National Party. In this they might be supported by lobbies and pressure groups overseas and possibly even by some local intellectuals.

Some of the arguments against a compromise which formally accommodates organised minorities have already been debated in the literature; a debate which the ANC might take as vindicating its position.

Heribert Adam, for example, considers that there is little need for a resolution to take account of white needs for political identity and some degree of autonomous mobilisation: "The National Party now represents the interests of the growing Afrikaner urban middle class and bureaucratic bourgeoisie. The logic of retaining BMWs and access to the spoils of multi-racial state capitalism is pitted against notions of self-determination in a self-reliant but poor racial state" (the latter referring to the CP position). Adam adds that he doubts whether the ANC would survive a compromise with the NP, that a compromise would amount to "an institutionalisation of invidious group affiliations" and would not be realisable in practice — "it may be better to let the conflict ripen".[2]

Many of Adam's reservations are endorsed by the influential director of the Institute for a Democratic Alternative for SA (Idasa), Alex Boraine, who concludes that "the problem in South Africa is not how to resolve two competing nationalisms — but how to build on the existing non-racial position". Boraine sees the idea of dualism in a compromise solution as "institutionalis(ing) the concept of division, and therefore of conflict".[3]

The arguments of Boraine are important, inasmuch as one might expect them to gather momentum among both English- and Afrikaans-speaking white intelligentsia. They illustrate that there could very well be a mobilisation of intellectual support for a position opposed to compromise with the National Party, not necessarily because of any opposition to an equitable compromise as such but because of very genuine idealism about what Boraine and others refer to as a transcendent South African nationalism or patriotism, which is obviously very desirable. Hence, a factor in the negotiation process, and one which could even undermine the prospects of compromise, is the prospect of a conflict between genuine idealism about a non-racial South Africa and the realities of a power-compromise, which will be at issue in the negotiation itself. The idealism will strengthen the prospect of the ANC resisting compromise, and justifying it not on the basis of its own power agenda, but on the basis of utterly respectable and compelling moral-philosophical analysis.

In fairness to the very legitimate non-racial transcendent position, one should ask whether one might not expect the National Party to withdraw from its negotiating position and accept an unqualified majoritarian outcome, seeking only a bill of individual rights as a safeguard for its constituency. This is essentially what the Democratic Turnhalle Alliance (DTA) in Namibia accepted when constitutional negotiation finally commenced. There are, however, signal differences between Namibia and South Africa, differences which cannot be debated in this chapter. This author, however, would argue that it might be very dangerous for a future resolution for influential pressure groups, and possibly external governments, to put pressure on De Klerk to forego his negotiating position, and for the following reasons.

Firstly, the growth in support for the Conservative Party has already been noted. More importantly, there is undoubtedly a proportion of National Party supporters who are prepared to endorse their government's negotiation agenda because they assume that De Klerk will be firm and tough in negotiation. If the government is perceived to be "climbing down" then the CP's opposition to the negotiation process will rise even further. De Klerk is committed in terms of election undertakings to test the outcome of negotiations among whites at least in a referendum. Quite frankly he may lose the referendum if he manifestly abrogates the commitments to his constituency with which he entered the process.

In this regard it is interesting to reflect on an item in the February 1990 nation-wide poll of Market and Opinion Research. De Klerk, with his authority and popularity as State President, has committed the government to abolishing the Separate Amenities Act. This would occur in a context in which virtually all private amenities are already desegregated — the Act relates to government and municipal amenities. This abolition would hardly be on a scale of sacrifice compared with the acceptance of an open numbers-based political system and one might have imagined that a clear majority of whites would endorse De Klerk's intentions. However, the poll showed that only 48 per cent of whites endorsed this relatively minor concession, and over four out of ten National Party voters and over three out of ten English-speakers would oppose De Klerk's intended legislative repeal. It is in the light of these kinds of facts that notions of discrediting De Klerk's negotiating agenda should be seen.

Secondly, the current actions of some of the ANC/MDM formations, while perfectly understandable in the light of pressures from their younger and more activist members and of dangers of being out-radicalised by the PAC, probably mean that rank-and-file whites are becoming increasingly less inclined to accept unqualified majoritarianism. Very recently there have been numerous protest marches and demonstrations at which the communist flag is openly flown and in which youthful activists parade with imitation AK-47 rifles. White motorists have been stoned on busy highways in the Transvaal. Mr Nelson Mandela repeatedly talks of the intended nationalisation of large industry and members of the United Democratic Front (UDF) have added that it should be without compensation. To his credit Mandela has attempted to reassure whites. In a recent speech in Bloemfontein, however, he probably did more harm than good by citing the Afrikaner "turncoat", communist Bram Fischer, as an example of the ANC's acceptance of Afrikaners.

One must expect that during this heady time of negotiation in a highly politicised society, there are bound to be excesses of political posturing on all sides. At the same time, however, it hardly seems plausible to expect whites to make large and sweeping concessions in the climate which is likely to accompany negotiations. Rank-and-file whites, like any other mass constituency, do not have the intellectual sophistication to discount political rhetoric and protests as being a necessary phase of release of political tensions.

A third problem revolves around what non-racialism means as it is envisaged by the ANC/MDM. One accepts that the sentiments are genuine. One also accepts that there is a perfectly legitimate and necessary goal of redressing black disadvantage. If the future "non-racial" government sets about intervening to achieve this redress rapidly by using state authority to restructure South African institutions (and there is no reason why it should not feel justified in doing so), the implications for whites could be fairly threatening.

To illustrate this point one may quote a small example. When F W de Klerk recently said that he was considering appointing a judicial commission to investigate the shooting by black and white policemen of black demonstrators in Sebokeng, Nelson Mandela welcomed it but expressed regret that the judge would not be black. The non-racial position, strictly speaking, is that the judge should be fair and balanced and of any colour. Heribert Adam refers to a possible future restructuring of the racial composition of the civil service (which employs 34 per cent of whites) as no more than "equitable". Many more examples could be given, but the point is that the "non-racialism", if it involves affirmative action, cannot be colour-blind. It involves reducing or retarding the advantages of one race to advance the position of another race.

One has to repeat the fact that the moral justifications are abundant. Nevertheless, whites who are in the affirmative action firing line will not see it as equitable non-racialism but as very much a racially based social engineering. In this sense it is impossible for the society, as it has become structured over centuries, to "have its non-racial cake and eat it". "Non-racialism" means one thing to blacks and another thing to those whites who do not have the money to escape its effects.

This is quite apart from the fact that non-racialism in the context of a seven to three African demographic preponderance automatically ensures a deference to the interests and symbolic needs of one of South Africa's racial categories. African interests need not be promoted. The logic of numbers will see to it as a matter of course.

To conclude on this issue, Heribert Adam may be right in saying that at present whites or Afrikaners, unlike the Northern Irish communities, are not a cohesive ethnic group with a strongly manifest primordially based solidarity rooted in myths of origin. He is wrong to assume, however, that this implies that they will willingly desegregate and blend into a society which they will perceive as pitted against their material, occupational and residential interests in the name of racial equity. To assume this is to endorse policies which will risk creating an emotional solidarity which does not exist yet among whites and Afrikaners.

In contrast to the Ulstermen or the Basques or the Tamils in Sri Lanka, South African whites as a collectivity do not have a fully blown myth of origin but they have an equivalent and equally powerful myth of mobilisation — "European standards". South African white ethnicity has been a self-reinforcing product of the very earliest contact between technologically advanced European settlers and technologically far less developed indigenous peoples, out of which a plural society emerged.

As Lieberson noted in 1961, some of the most intractable inter-group conflicts flow out of deep historic conditions of the first encounter.[4] Today, apartheid and the universal tendencies for social and economic advantage to be perpetuated through the family system have ensured that, in terms of broad proportions, the relative educational-technological disadvantage of blacks is almost as great as it was in the 17th and 18th centuries. There has not been sufficient black occupational mobility to break the apparently valid stereotypes, as it were, and this is in some measure due to economic constraints imposed by sanctions and political instability.

While white social identity may not be primordial but rather an "instrumental" unity, a kind of popular class advantage, most whites nevertheless adhere to the myth that there is some kind of very basic contrast between First World and Third World. They see their middle-class privilege as the manifestation of their origin as Europeans, protecting certain "standards" in a sea of Third World conditions.

This is not to defend the white myth. It is simply to say that whether the roots of white identity are primordial or instrumental is not the issue. In politics perception is everything and whites perceive themselves to have a social identity, latently as powerful as the commitments of true ethnics, as the protracted struggle in the old Rhodesia showed. The whites, without contemplating its possible shallowness (Afrikaans ethnicity excluded), have a myth of lifestyle which, because of its coincidence with colour, is tantamount to the identity formation in ethnically divided societies. Whites will compromise, they will open the formerly closed group boundaries and they will share resources and power, but they probably will not put their myth of having special standards to protect on the line.

These needs, which have been referred to as "white" perceptions, are in a very powerful sense "backed" by Afrikaner ethnicity. It is a truism that nationalism becomes manifest under perceived threat. For four decades Afrikaans identity has not been threatened. Afrikaners have been secure within the sphere of their effective control, by virtue of demographic advantage, of wider white racial mobilisation. Because white domination, not Afrikaner domination, has been under attack, Afrikaner identity has had the luxury of subsiding to a level of private commitments and latent values not relevant to immediate political challenges and threats.

If the shelter of the structure of white domination were to be stripped away, and Afrikaners' cultural symbols and their collective sense of "place" in the society were to be an issue, virtually all comparative and historical precedent would suggest that Afrikaans "nationalism" would come to the fore once again. As Kenneth Boulding observes, "nations are the creation not of their historians but of their enemies".[5]

In accounting for the persistence of apartheid, however, the underlying and currently largely understated Afrikaans ethnic orientation has permeated the commitment to racial minority group coherence and self-determination, giving it a consistency which is not present among English-speakers. A very recent example, albeit possibly mobilised by the CP, is the rejection by Afrikaans parent associations throughout the Cape Province of a new government policy of allowing segregated white state schools a local option to desegregate if a very substantial majority of parents agree. Even this cautious reform was felt to threaten an ethnic tradition.

There is, of course, a very easy way of resolving the problem at a moral and intellectual level. Since it is not "equitable" and results in part from apartheid, one can discredit this mobilised white interest and join the clamour for its destruction. This, after all, is what most analysts do without much hesitation.

The problem is that whites are willing to endorse negotiation precisely because they consider that the National Party will protect what is sufficient for them to protect their myth. Without the possibility of compromise between these minority interests and legitimate majority interests, negotiation would not be on the cards.

Broader thoughts on strategy

Before drawing final conclusions there is one fairly prevalent notion that has to be addressed. This is the view that one can afford to put more pressure on the government than on the ANC as regards their basic constitutional positions because the government has no options. The argument would be that it has to negotiate under almost any terms in order to resolve crises which have arisen because of sanctions and because of internal black resistance. Certainly the government is under considerable pressure but to assume that it has no options could be very misleading. To understand the parameters of the present transition, one has to consider a number of realities.

Firstly, since assuming the leadership of the National Party government, De Klerk has made his unexpected moves to liberalise the political process in South Africa from a position of (initial) strength. South African export performance had surged despite sanctions. The ANC, having been deprived of its Angolan training camps by the agreement in the Namibia settlement, must have lost confidence in the traditional armed struggle. Apart from Natal, at the time the new climate emerged internal protest and revolutionary dissidence was at a low ebb, ground down by the pervasive network of security management established in the latter period of P W Botha's "imperial presidency".[6] Nevertheless, the climate of protest and dissent persisted, clearly indicating that repression was an insufficient means to address the basic political problem. Reform without negotiation was an option, however.

The collapse of communism in Eastern Europe and its decay in the Soviet Union and China made irrelevant all the domino theories in international relations and the consistent analyses by the South African security establishment that the ANC would be used by the Soviet

Union to establish a presence in Southern Africa.

Politics is all about timing and the time was ripe for De Klerk, with the confidence of a reasonably equitable outcome in Namibia and of a critical but conditionally supportive West, to attempt to secure a settlement under conditions which would enhance his chances of preserving the crucial political interests of his constituency.

Conceivably, one of the greatest strategic mistakes that can be made is to ignore these circumstances and assume that the transition was commenced by the government from a position of desperation and weakness. This is not to say that De Klerk did not have problems to resolve. The 1989 election produced a resurgence of growth in opposition to his left and right, which made it clear that crucial choices had to be made. Furthermore, the challenge of restoring international credit flow to a capital-starved economy with rapidly mounting unemployment was a powerful lever on De Klerk. These factors notwithstanding, his strategic position was more favourable than it could be calculated to be ever again. The government considered that it could risk the effects of releasing Mandela and unbanning the ANC, on the assumption that the ANC, despite probable hesitation and contradictions, would have to negotiate within a framework compatible with the bottom-lines of the government and its constituency.

Secondly, the De Klerk administration is by no means the last bulwark of white political resistance. As already argued, before De Klerk's very recent moves, which have disconcerted even some of his own supporters, the right-wing opposition is estimated to have grown to contain at least some 38 per cent of white support. If circumstances were to arise, such as mounting civil unrest or destabilising world reaction, in which De Klerk loses his nerve and begins to make large, forced concessions, the right-wing opposition party, the Conservative Party, acting through its formidable support in the white civil service and security forces and by inducing uncertainty in more conservative local National Party organisations, could force an election and take over (with 42–44 per cent support in the present constituency delimitation, they could gain an overall majority). If no election is allowed a military cum bureaucratic coup is not to be excluded. A much more brutal articulation of white interests waits in the wings.

Thirdly, the attempts which are likely to be made by the ANC/MDM to increase their leverage in the transition must almost inevitably include some attempt to emulate the model of mass dissent which has worked so well in Eastern Europe. The model is simply too dramatic and too topical not to be tried. Yet in Eastern Europe it succeeded fundamentally because the communist regimes had lost accountability to their constituencies and their bureaucracies were themselves in a state of political fragmentation.

These conditions do not even remotely apply to South Africa. The effect of attempts to mobilise vast crowds in an assault on the "Berlin Wall" of apartheid might cause paroxysms of anxiety and an increase in emigration among uncertain white liberals, but it is also likely to induce a fear which will stiffen resistance in the mobilised centre-conservative constituencies of the National Party and Conservative Party.

It is essential, therefore, not to underrate the conditions which made it possible for De Klerk to move boldly. If these conditions are reversed, in an attempt by the ANC or PAC or the international community to mount pressure for far-reaching concessions or to obtain a quick result, the government could revert to its own historically familiar mode of tough resistance, this time accompanied by co-optive socio-economic reforms. The grip of history might have been relaxed but its tendons are ever ready to tighten again.

The conclusions from these arguments must be predictable. The analysis has suggested that the kind of compromise which will flow from what will be tough negotiations between mainly the ANC and the government is probably the best scenario which South Africa is realistically likely to achieve.

One can immediately anticipate the "what about the workers?" objection. Why should the internationally endorsed commitment of the ANC/MDM to a non-racial democracy be discarded on the basis of morally dubious strategic considerations? What about the masses? Why should the white constituency count more heavily than the black mass constituency? These are good questions and there is no earthly reason why the white constituency should be more salient than the black constituency. The short answer, however, is that it need not be more salient.

A number of empirical investigations have shown that the mass of ordinary rank-and-file blacks, excluding activists within the MDM and PAC, appear perfectly prepared to accept a power-sharing compromise. A 1987 study by the Emnid Institute of West Germany on behalf of the Deutsche Afrika Stiftung among 1 004 black coalminers showed that only 27 per cent endorsed "majority rule" in preference to balanced ethnic power-sharing.

Another study, by Nic Rhoodie and M P Couper, showed that no less than 75 per cent of a sample of 1 338 urban blacks in the PWV region in 1986 considered a "government consisting of all four population groups without domination by one group over other groups" to be very good or good; a higher proportion than that endorsing a majority-based government.[7]

A study by the Argus Group's Marketing and Media Research in October 1988 among 382 blacks in the PWV, well over 60 per cent of whom supported the ANC, showed that only some 33 per cent preferred a "single, mixed-race parliament with the majority party in control" to other options involving either group-based power-sharing or decentralised government.

The point in all this is that while the National Party has a fairly limited potential white support-base to its left, the ANC, whether it realises it or not, would increase its committed support if it were to turn to the black middle ground on a ticket of meaningful power-sharing, job creation, urban development and economic growth. It need not be confined by the interest articulation on the left wing of mass black opinion.

There are many other factors beyond those discussed which are relevant to a successful resolution, the nature of the future economy being among the more important. This analysis, however, will have to suffice with the issues addressed. These suggest that the slate of history, no matter how tragic it has been in the past, cannot be wiped clean in the first outcome of South Africa's process of national reconciliation. Apartheid's dismal legacy cannot simply be engineered or swept away. A bridge to a non-racial democratic future is achievable, however, and it might be the wisest route to follow.

ENDNOTES

1. Francis Fukuyama: "The End of History?", *The National Interest*, Summer 1989, pp3-18.
2. Heribert Adam: "The Polish Path?", *Southern African Review of Books*, October/November 1989, pp3-5.
3. Alex Boraine: "Pitfalls of Focusing on Groups in South Africa", *Democracy in Action*, August 1989, pp2-3.
4. Stanley Lieberson: "A Societal Theory of Race and Ethnic Relations, *American Sociological Review*, Vol 26, 1961.
5. Kenneth Boulding: *The Image*, Ann Arbour, Michigan University Press, 1956, p114.
6. Brian Pottinger: *The Imperial Presidency*, Johannesburg, Southern Publishers, 1988.
7. D van Vuuren *et al* (eds): *South African Election*, Pinetown, Owen Burgess, 1987, p227.

15 The high price of peace

BERNARD CRICK

This is a speculation in applied political thought. It is not about the empirical nature of the three problems themselves but about how one thinks about problems that appear to be insoluble because, symptomatically, normal politics breaks down when protagonists pursue policies which they say admit of no compromise and which are believed to be totally exclusive and mutually contradictory, at least as formulated by the main antagonists. Each of the three problems before us have in common that the normal democratic principles of majority rule are part of the problem as well as part of any possible resolution. So my object is not to propose the best solution nor attempt to adjudicate between rival claims, but rather to explore how best to perceive such problems, how best to approach them. The English country story has an almost Hassidic wisdom about it. "How do I get to Biddecombe?" "I wouldn't start from here if I were you." So I want to make a mild humanistic protest against the prevalence of restating the arguments of the protagonists as if they are possible solutions, and against intellectuals and concerned communicators joining and taking sides, as if they were lawyers in court in a common law advocacy system, rather than considering how best to approach the attainment of acceptable peace or conciliation.

Much good reporting and commentary is also marred not so much by clear partisanship, but by a tendency to see and present the problem only in terms of the day-to-day recurrent injustices, especially the violent injustices, committed by one side or the other. This creates a danger of responding passionately to symptoms and not thoughtfully to causes; the public in Western democracies outside our three troubled areas must sometimes wonder what the underlying conflicts are about, why are they doing this? The "atrocities of the terrorists" and the "excesses of the security forces" are events which are themselves part of the propaganda battle, and often manufactured or provoked for that purpose. And all factual balance sheets of incidents and atrocities are highly subjective in their assumptions. While it is extremely important to ask activists what they want, and to listen carefully and empathetically to nuances in their replies, it is just as important, sometimes more so, to stand back from the struggle and to ask close observers (like journalists, churchmen and businessmen, for instance) what they think is possible. One should try to establish a way of looking at some extreme problems that can do justice to both sides when both sides may have, in many senses, a plausible, and to the individuals involved even a just case.

The problem of the starting point

In my *In Defence of Politics* I spoke of making some old platitudes pregnant: that politics is the conciliation of naturally different interests, whether these interests are seen as material or moral, usually both. I argued that all advanced societies contain a variety of values and interests, and that conflicts are best settled politically. Politics, in this old Aristotelean sense, is both historically and logically prior to democracy: again, in some situations the simple application of a democratic majority rule to resolve conflicts can in fact increase the conflict, make political resolutions more difficult. This is not an argument against democracy of course, only against seeing it as a single universal and overriding value (Tocqueville and J S Mill have said everything that needs to be said on this). Not all societies are organised and governed according to political principles. Some see politics as inherently disruptive of social order and politicians as, in Goebbel's words (that could have been Stalin's, or Pinochet's today) "existing to perpetuate problems not to solve them". Political rule, I said, existed before democratic government, and some dictatorships are genuinely popular, resting on majority support and the stronger for it.

I argued in my *In Defence* that political principles or the pre-conditions for politics (leaning on old Aristotle against the over-sophistication of modern social science, whether Marxist or American-behavioural) have two aspects, a sociological and a moral. The sociological one is that societies are seen to be complex, inherently pluralistic even if and when (hopefully) all unjust discriminations of class, race and gender vanish or diminish. The moral was that it is normally better to conciliate differing interests than to coerce and oppress them perpetually, or to seek to remove them without their consent. Certainly much political behaviour is prudential, but always in some moral context: there are some compromises we think it wrong to make, and some possible ways of coercion or even defence which we think are too cruel as well as too uncertain. A nuclear first strike, for example, even against a non-nuclear power, could not reasonably be called political behaviour. Hannah Arendt was wiser than Clauswitz and Kissinger when she said that violence is the breakdown of politics not its "continuance by other means". But politics can break down, and there are circumstances, of course, when violence is justifiable. In both classical political philosophy and in Christian theology there are, for instance, definitions of justified rebellion and of the duty of tyrannicide. Arendt says, however, that a necessary condition for violence to be justifiable is that its object is immediate and precise.

Political prudence calls for good judgement and rational calculation in relation to the external world which is not always present when a community feels, rightly or wrongly, that its most precious distinctive values, perhaps its very identity, are under threat. Would real economic pressures and sanctions on the Unionists or the Afrikaners, even the present government of Israel, make them more willing or less willing to compromise, more or less paranoic and self-sufficient? This is very problematic. Both liberal economists and Marxist socialists under-estimate the strength of desire for cultural autonomy in each of these situations, however economically irrational that can be.

So politics is a conciliating activity or process, not a set of substantive rules needing (as unthinking politicians say when trying to be thought-ful) agreement or consensus about fundamental values or even the rule of law. Politics can change laws peacefully and find paths of compromise amid differing values so long as there is a broad consensus about procedures. There is more agreement in societies like the United States and the member states of the European Community about *how* to conduct disputes than about the *ends* of policy. One can sensibly invoke the concept of procedural values: at least a minimal respect by political activists for freedom, toleration, respect for truth, as well as empathy and a willingness to settle disputes by public discussion.

But what happens when none of these prior conditions for politics exist, when procedural values themselves are hotly contested? The three cases chosen, Northern Ireland, Israel/Palestine and South Africa are at least intellectually interesting to the political thinker, even if there were not moral and practical reasons for our interest. Political leaders in each of these areas appear either to fail to perceive the complexity of the interests to be satisfied (as if their perceptions are affected by the intensity of the conflicts), or believe that the very safety of the state or the survival of their people, as the case may be, is threatened by the political articulation of the rival interests. They then conceive it as their moral duty not to compromise but to oppress, eradicate or simply endure. They see their opponents as beyond political persuasion. They see them as threatening not merely their basic interests and beliefs but all that they are, their very being, their precise human identity: thus as literally beyond reason, like another order of beings. *All* political opposition becomes seen as inherently subversive. Such terrible fears can arise from religious, ethnic, national or class prejudice: the others are not like us as a people and never can be, and nor have we anything to learn from them. And this is often a reciprocal relationship, a self-fulfilling prophecy. "Treat people like rats," said Arthur Koestler, "and they will soon learn to bite your fingers off."

When large numbers of people are treated as enemies of the state and its possessors, then they may also attribute to themselves a complete and artificial unity. This may be false culturally and hard to sustain politically (blacks and whites; Arabs and Jews; Irish and Brits: "they are/we are all the same"), hard to sustain without repression within their own ranks. Nationalism for the time of struggle is so unifying and intoxicating that it has often tempted leaders elsewhere to continue the struggle even after victory, or victory enough. There are cases when some of the disenfranchised and dispossessed are as prejudiced, violent and irreconcilable as the rulers. George Orwell once said that after becoming anti-imperialist it took him twenty years more to realise that "the oppressed are not always right". (He didn't mean they were not right in not wanting to be oppressed; but they were not right, or necessarily specially admirable in everything else. Western friends of the oppressed can be embarrassing idealisers; I like Mark Twain's humanism: "God damn the Jews; they are as bad as the rest of us.") Orwell's remark is a salutory reminder. But the common-sense moral view is that rulers have more opportunity for reform and therefore carry a greater moral responsibility for failure to act effectively against endemic disorder, poverty, injustice and (the most telling monitor of all) differential life chances between different groups. As many revolutions take place because governments break down through the ineptitude of conservatives as through the prophecies, schemes and courage of revolutionaries.

In other words, what happens when normal politics breaks down, or remains minimal and locked in a dominant community which is either not accepted by a clear majority of the inhabitants (as in South Africa) or by minorities who consider themselves rightfully a majority but frustrated by boundaries unjustly and arbitrarily drawn (as in Ireland and Palestine)? One odd feature in common of the three actual regimes in these cases is that none of them are pure autocracies but that each has some kind of working parliamentary institutions, indeed a vigorous and by no means superficial political life within the dominant community. Nonetheless, in each case the existing institutions do not appear to furnish a mutually acceptable framework for the resolution of conflicts; rather the discontented see the existing institutions as part of the problem. They do not merely want, like most American blacks, a fair share of conventional goods and opportunities and to be treated as ordinary Americans. They want fundamental changes in institutions as well as in the balance of powers between communities. The existing institutions are not seen by all as even an acceptable framework for negotiation, quite apart from refusal to negotiate at all.

I call the three problems "insoluble" for two formal reasons: (i) that no internal solution likely to guarantee peace can possibly satisfy the announced principles of the main disputants; and (ii) that any external imposed solution or enforced adjudication is likely to strengthen the desperation and self-righteousness of the threatened group (I first heard the phrase "laager mentality" in Ulster, long before I heard it in South Africa; and Israel knows "settler mentality" and "uncompromising defence of the settlements" too). No one can win in terms of their expressed objectives, victory would be defeat, pyrrhic at least. Indeed almost any gain for one side is immediately pictured as a serious loss for the other. Hence typically there emerges an ideology and personal ethic of defiant heroism: "sacrifice for the cause", "he dies for his people", "the blood sacrifice", "better to die fighting than live in subjection" and so on. The thinking of Zionist and PLO, Unionist and IRA, and Broederbond and ANC have obvious similarities in these respects as well as differences (indeed each, in some respects, inspires the other). Each governing community claims a unique religious identity, even destiny, as well as an ethnic peculiarity. And both opponents and supporters of the IRA, the PLO and the ANC can see *some significant similarities between each other*.

In all three cases I have become almost obsessed with the wisdom of assuming an equal justice to the broad case put by *both* regime and rebels — *if* one knows something of their history *and* grants their basic premises of argument. This is painful but may be a better starting point for any possible conciliation than trying to decide *who is right*, whether by historical argument (who got where first); by a balance sheet of rights and wrongs (by analogy to better legal and political practices elsewhere);

or by utilitarian argument (who makes or could make the best use of "the land" or "our land"). Any empathy is painful and fraught with misunderstandings because some government supporters, say, are so unused to empathetic understanding, especially from foreign observers, that they often mistake it for moral support. And those in opposition similarly regard any scepticism about their means and their slogans, let alone about their ends, as treachery, weakness or collaboration with the enemy. Overseas supporters (who have adopted them as a cause) are often especially virulent: "So you went there! You don't believe in *total* support for the boycott?"

Albert Camus once raged against his fellow intellectuals in France who thought it their duty to take sides in the Algerian War and to provide justifications for "the right to slaughter and mutilate", each relying for their justifications mostly "on the other's crime".

> The role of the intellectuals cannot be . . . to excuse from a distance one of the violences and condemn the others. This has the double result of enraging the violent group that is condemned and encouraging to greater violence the violent group that is exonerated. If they do not join the combatants themselves, their role (less spectacular, to be sure!) must be merely to strive for pacification so that reason can have a chance.

Some may follow my obsession or painful empathy into Israel/Palestine and Ulster/ Ireland but may find it hard to follow, or think it quite perverse, when applied to South Africa. Certainly there are clear arguments why racial prejudice is morally more unjust than even religious bigotry, or the conceits and constraints of class; and the numbers excluded from the franchise in South Africa are impressive by any standard, especially in a state that claims to be parliamentary and democratic even in part. But in each of these cases, South Africa not least, a large part of the basic problem is the desperation of historic communities to preserve their identity (originally Afrikaners against the British, quite as much as both the indigenous and migrant peoples against the oppression of both, sometimes against each other).

To remove all violations of basic individual human rights is a necessary condition for any stable and just society in South Africa, but it is not a sufficient condition. Quite apart from the problem of poverty, so many — surely most? — individuals in South Africa find their human identity in being a member of a specific community. And each of these communities can appear to threaten the others (indeed does), hence threatening each other's most basic sense of identity, not simply their economic interests or acquired beliefs in democratic (or other) political institutions. To say, as the ANC, for instance, has often appeared to say (more often their supporters outside Southern Africa), that "one person one vote" constitutes a solution, is only to point to part of the problem, just as it does in Northern Ireland and in Israel where the use that the majority make of their power is part of the problem and not by itself a sufficient solution.

In such situations neither side can gain victory except at a price normally too great for men and women to contemplate, nor can any established juridical or democratic criteria be applied to reach any mutually acceptable result. The breakdown of normal politics creates the need for a kind of politics that can find guarantees for continuing communal identities as well as common citizenship. Few politicians, few thinkers even, have even begun to think of the implications of a pluralistic democracy. What constitutes a community within a state with different legal rights, whether religious, ethnic or territorial? And if to some degree or other such communities may enforce their own rules, how far should the state not merely monitor those rules but insist that individuals (especially adolescents and women) are not imprisoned in those communities? Legal apartheid is an abomination but in most countries intermarriage between people of different communities (where community involves a strong ethnic or religious loyalty) is still not easy; and in most countries the tension between community values and individual rights strikes women particularly hard.

Suppose leaders talked a language of generally *acceptable* resolutions rather than democratically *agreeable* solutions? The price for any such accommodations would be painfully high in each of the three cases, in terms of previously announced "unnegotiable positions" and what communities and their existing leaders conceive to be their rights, both against others and to manage their own internal affairs without interference. And in each case it will involve talking seriously and respectfully to people who have done "unforgivable things". But paying a high price has to be faced at some stage if the alternative in each case would be (as is so often threatened, to try to bring the unruly to the negotiating table) chaos, anarchy or the breakdown of the social order, or even more likely, a sad and sordid continuation of present levels of instability, violence, tension and injustice.

Any longer and empirical study of the basic problems of instability and violence in these three areas would show their complexity and might stress the *differences* there are within what only their opponents and foreign supporters see as a monolithic Catholic or Protestant, Arab or

Israeli, black or white camp. (Activists on the ground are much more realistic, indeed commonly talk as much about divisions in their own camp as about the nature and iniquities of the regime.) Nonetheless, in each case the fate of moderates who have attempted to use these differences within their communities to build a political base for compromise is not encouraging. A community, when it fears that its very existence is threatened, rallies to leaders who they can be sure will defend it utterly and by any means, even if those leaders' beliefs and means (as with Ian Paisley) are not normally acceptable. Such leaders are in many respects unrepresentative except as militant defenders. Advocates of moderation are too often outflanked.

A longer study should make few direct comparisons between the circumstances of the three areas and promise no easy answers. There are none. Each is historically and sociologically unique. The common factors are found not in the empirical facts but in how we perceive these problems and in how we try to think about them. Perhaps it is their very complexity or perhaps a natural impatience at their bloodiness and violence that tempts people into specious comparisons and general "solutions". These solutions are usually a list of "self-evident" general principles so presented as to lead deductively to a card castle of detailed institutional proposals. It is hardly too harsh to say that when anyone cries "I've a solution", one knows one is dealing with a fanatic, an innocent or a crank. "Patience and time," said General Kunotzov in *War and Peace*, contemplating his young officers eager for immediate battle with Napoleon.

Strictly speaking, and there is need to speak strictly, only *puzzles* have unique solutions. *Problems* (especially political problems) can either be said to have many possible solutions, or no solutions, only resolutions, settlements, compromises or even ameliorations; none perfect, but several, perhaps many, ranging from the more or less agreeable to the more or less acceptable. Real differences of interest may continue but they can, with will, skill, resources and good fortune, sometimes be made or become less intense, more peaceful. But such resolutions will rarely be what anyone deeply involved would have rationally agreed to in advance of events and negotiations. So I will try to offer a framework for thinking about such conflicts that offers neither too much nor too little. Impatience in remedying injustice is the mirror image of delay in grasping the basic problems of deeply suspicious if not hostile communities living in the same territory and each claiming mastery of it.

The high price of peace

My moral is not peace at any price but the terribly high price and unpredictability of victory if it comes to winner-take-all. The price of any negotiated peace in such situations is high enough and painful enough, in terms of the present expectations of the activists, if ever the main actors come to want peace more than victory or the glory of continual sacred struggle or resistance. A desire for peace can come through exhaustion and war weariness (as in Ireland in 1919 and 1920), through moral conviction amid a diminished fear of losing communal identity, or through fear of losing everything. When these conditions begin to appear, then, and only then, outside aid and pressure can be of much help, indeed sometimes crucial; until then, each of the three situations suggests, external interventions can actually increase fears and prove counter-productive. Inevitably the form of such interventions owes far more to the nature of the domestic politics of the donors than to any objective appraisal of the needs of the receivers.

The highest price of peace is likely to be giving up hopes of victory, a realisation of the limits of power. Israel must not, indeed cannot, be displaced but equally it cannot govern areas which are predominantly Arab; nor can it deny to Arabs in these areas the right to be governed how they choose and by whom they choose. Northern Ireland cannot be governed peaceably from either Dublin or London alone. The province (and it is a province not a state, nor even a normal part of a state, both in relation to Ireland and to Great Britain) inherently faces both ways. Few people in the North act as if their identity were exclusively Irish or exclusively British: but leaders cannot take this reality into their policies because of exclusivist fears stirred up by their rivals. Populist leaders of ethnic or religiously identified communities are necessarily more responsive to their peoples. The Anglo-Irish Agreement points a broad way forward towards institutions of joint responsibility, dual citizenship and bi-communal equality; but its effectiveness is hampered by the exclusion of one community in its formulation. In South Africa the compromises will be between communities: the ending of apartheid and the freeing of labour in a more rational capitalist market could be a step forward which both sides could support, though expecting different consequences; but the black and coloured communities are so disadvantaged economically that deliberate economic and social reforms by the state will be needed, though through time.

It is important to consider the nature of time in society. What kind of changes (whether of income, education, health, the structure of the economy or basic attitudes) take what kind of investment lead-time? "Rome was not built in a day" or, as William Blake said, "eternity loves the products of time". A real conviction may grow that things must change or at least cannot go on as they are. But opposition leaders must realise that real changes in attitudes among a ruling class and real changes in skills and educational attainment among the dispossessed are a matter of decades, sometimes generations, not of a parliamentary session, a "night of long knives" or any mythical "first hundred days". Revolutions are long processes not quick events. The distinction between revolution and evolution is not absolute. The reforming politician with a sense of time (like Lincoln) is a statesman indeed; the conservationist politician who uses time as an excuse for delay is, indeed, a timeserver. No modern state can now turn back the rising tide of expectations among the dispossessed within its boundaries, nor demands for free movement and expression: for these are externally triggered, not necessarily by deliberate pressure but simply by communications and involuntary example. Both man and monkey are mimetic.

The changing international situation not merely makes the deliberate stirring of conflicts in the Middle East and Southern Africa by the US and the Soviet Union less likely, but creates the possibility of a joint interest in stability — not perfect justice or agreement, but things acceptable.

I have stated two general reasons why each of these problems can seem insoluble. A third contingent one will emerge in looking at the three areas separately: the virtual impossibility of successful armed rebellion against each of the governments concerned. But also there are severe political, moral and international limitations on the force that each government could use against internal subversion: hence the relative ease of spasmodic violence and terrorism, but also the real aim of *most* of the terrorism in Israel/Palestine and in South Africa is not to bring down a regime but to force it to the negotiating table. In both cases, terrorism is unlikely to do so on its own: it stiffens resolve and heightens fears in the regime, but when used tactically in combination with civil demonstrations and strikes and foreign pressure, is more likely to succeed — and, seen in that light, is in principle more justifiable. At least it is then more absurd to demand that the oppressed forswear all violence as a condition of negotiations. In such circumstances negotiations offered by the government are unlikely to be genuine unless constrained, that is, that both parties in any real negotiation must believe that the consequences of the negotiations breaking down will lead to a worse situation than before.

A vital difference emerges, however, between those terrorist groups who have political aims, or more precisely those who have such stated aims — however vague, idealistic or impractical — yet are capable of acting politically, and those whose behaviour is stridently anti-political, those to whom *the struggle* and heroic defiance have become a way of life, ends in themselves. The PLO and the IRA seem to have real divisions between politicals and militants. The ANC seem more tactical in their appraisal of means and ends. The politicals can be seen as extremists having some genuine claim to be representative of, as well as leaders of, a mass popular *movement*, but the militants while claiming to represent *their* people, in fact act like a *sect*. A sect is internally motivated. Any sect is difficult enough to deal or bargain with, and many sects, mutual rivals for leadership of a movement, are almost impossible to deal with. (The growth of Islamic fundamentalism proves a common problem to Israel, Egypt and Syria.) The PLO and the ANC are not divided into clearly institutionalised political and military wings, as each exhibits shifting and tactical moves back and forth from one to the other, and with few individual members always wholly one or the other.

However, the stalemate of subversive terror versus state terror, neither able to win but neither able to escape the other, can sometimes create a qualified general ground for hope: out of the very eye of the storm can come war weariness and the meeting face to face of key or honoured leaders on both sides.

Three general perspectives or maxims of political thought will emerge:

(i)   Concerning the nature of state power — for all its abuses in each case, its limitations are also clear: the inability of the governments to silence their opponents. There is a vital difference between power as unchallengeability (no one else can do it, even if I can't) and power as the ability to carry out a premeditated intention (getting things done). Some governments can't be overthrown but they can neither prevent unrest by force nor carry their followers into reforms. So the state's power to undertake reforms depends on its ability to mobilise and convince the alleged beneficiaries of reform as well as its own supporters. That is the very difficulty. Power to be powerful needs redefining as legitimate authority. To get such authority, it is not foolish or empty to say, governments actually have to act reasonably legitimately. In transitional situations it is difficult, indeed, to treat yesterday's gunmen as responsible opposition leaders or even colleagues, and to face the possibility of losing power in elections in some not too distant tomorrow. But the price, in terms of perpetual uncertainty and debilitating insecurity, of refusing to negotiate can come to seem even higher — which is another reason why the sometimes well meant demand that oppositions in such situations should totally renounce the tactic of violence is both unrealistic and unwise. For the price the regime has to pay is so unwelcome and begrudged that the pressure has to be kept up or else the government may lapse back or simply play for time meaninglessly, simply putting off harsh compromise or half-genuine attempts at reconciliation to the next man's turn in office.

(ii) There can never be peace in Ulster and in Israel/Palestine while the idea is held to that for every nation there must be a state, or that the state or national parliament, even, need be fully sovereign. For entrenched communal guarantees seem at the heart of each of these problems, whether protected by courts or international treaties. In both Israel/Palestine and Ireland even better demographic boundaries would still leave many bi-cultural areas and in each one a formidable territorial problem: the Catholic areas of Belfast and Arab Jerusalem. Some areas of the world do not fit the normal expectations based on the European experience of state formation since the 17th century and the almost universal emulation or imposition of it elsewhere. Political power is not indivisible, it can perfectly well be divided and limited. The Treaty of Rome is a clear enough example. Elements of federalism and federal thinking are almost inevitable in any lasting settlements, or frameworks for peaceable evolution by consent in these areas; only a federalism that will have to be communal as much as territorial. Pluralism was once a philosophy of the state critical of the theory of sovereignty. Some groups, it was argued, do have a relative autonomy that is beyond the power of the state. "All power is federal," said the young Laski. That is doubtful. But some power to be effective can only be federal. Sovereignty thinking, like simple majoritarian democratic thinking, can again be part of the problem more than the dominant part in any specification for a solution or resolution. The claim of sovereignty often precludes any genuine constitutional restraints, hence makes any negotiation still depend on the good will (which may be wholly lacking) of the majority or the more powerful minority. But two or more nations can inhabit one state and preserve their identities. We in the United Kingdom should sometimes look under our noses, as well as to Canada and Belgium and so on.

(iii)   The basic philosophic notion of a free action as an unpredictable action (one not entailed or determined by any necessary circumstances) needs relating to the very nature of political negotiation, bargaining and compromise. (Only in efficient autocracies are political outcomes reasonably predictable, and efficient autocracies are getting hard to find.) Certainly what the leaders of rival groups in the three areas say they each want as their unnegotiable objectives are relatively unlikely to occur. So if the real object is not revolution or victory but is to bring to the table the reluctant party or parties, then it is perfectly sensible, while not going into negotiations without preparation, yet to be uncertain or open-minded about what will emerge. No one will convert the other but in the nature of the problems and of negotiation, the precise outcomes are unpredictable. Again, only puzzles have unique solutions; problems can be resolved to tolerable levels, if never wholly removed, by negotiation in many different ways. Safeguards, privileges, rewards, opportunities and legal and social rights can all take many forms.

By the mid-1980s, for instance, it had become clear to the British and Irish governments that to some extent they had to act jointly, "in collusion" even (as the rival irreconcilables and invincibles saw it), over their common problem of Northern Ireland, even if the problem took a different form to each. They were uncertain what framework to advocate, what might work and prove acceptable to their rather different constituencies. Neither side brought a draft of the Anglo-Irish Agreement's Intergovernmental Council with them to the table. It emerged from negotiation, or some might say (but with hindsight) from the logic of the situation. It was not an instrument of government but rather emerged as a reasonably stable forum for consultation and continuing negotiation. Very probably it points, as some fear and others hope, towards some kind of evolving joint authority, joint citizenship and some working doctrine of communal equality. But to some on both sides, any negotiation was selling out. So sectarians commonly refuse to negotiate or try to wreck negotiations, until their own followers either grow weary of them and rebel, or more often, alas, drop out of politics entirely, leaving the fanatics and the "incorruptibles" unrestrained in speaking for them.

"Oppression driveth a wise man mad," says *Ecclesiasticus*. Happily this is not always true, sometimes it brings out the best in human beings.

I can find some hope not just in new ideas about constitutional arrangements but in what I found in South Africans under the ban — qualities not always associated with normal politicians: courage, dignity, tactical skill, understanding of their oppressors even, and an ability to state the essence of their case cogently, that is briefly and with absolute relevance as if genuinely to persuade. But if the other side are not so good at persuading, and often, like elsewhere, seem trapped in their own history, while it makes it more difficult to understand their true case and what are their minimal demands, yet it makes it more urgent to discover them and take them seriously.

So this will have been a plea not to offer either unqualified and uncritical support (though some are more worthy, indeed, of qualified support than others) or glib solutions for these very difficult problems which must, indeed, be tackled for moral as well as prudential reasons. There is a danger that outsiders can indulge their own emotions with either no effect on the problems or bad ones. Some lurid examples could be given. There is a need for both influence and critical judgement (in Kant's sense) from outsiders, or from insiders forcing themselves mentally to act like dispassionate outsiders: spectators do see more of the game and informed spectators can influence, not a particular game, but the way it is played in general. But such influence is at best indirect: the observer should neither volunteer for the front, as it were, nor make high-brow propaganda for the better cause, but should try to suggest the pre-conditions for negotiations between the actual parties involved. Because these situations are real and terrible problems and not perverse puzzles made for our solving, outcomes can vary and will always disappoint, somewhat; there are no perfect or unique solutions. Concerned persons should be less concerned to predict and support most favoured outcomes and more willing to imagine and to accept many possible and often quite unexpected outcomes.

Postscript

If peace and justice depend on the pre-condition of diminishing central-ist sovereignty thinking for pluralistic thinking, then that pre-condition, in order to be institutionalised through negotiation, must take a more legalistic form than has been common of late in European political thought, scarcely at all in the new nations. Negotiations themselves will be about constitutional devices and the only possible outcome will be a just, or radically more just and acceptable set of constitutional arrangements. Through these arrangements substantive issues of social policy would, in the near or middle future, be settled; but not in the constitution itself. The most that can and should be hoped for in negotiation is procedural consensus: agreement on rules by which substantive problems can, within a framework of government, be resolved.

Put it this way. Remember that the thirteen colonies in America that revolted against Great Britain found a common bond in a constitution they created themselves, long before there was a sense of an American nation. Constitutionalism came first. The first stages of the French revolution were constitutionalist, heavily influenced by the American example and by the writings of the Anglo-American republicans and constitutionalists and their French disciples. Only when foreign warfare and internal violence became general did nationalism stir and become stirred. The predominant reformist and emancipatory idea everywhere in the early and mid-19th century was "give us a just constitution" (1848!) and the American anti-absolutist slogan, "a government of laws and not of men". General legal rules rather than paternal particularism, or "each case on its merits", were in part a response to particular difficulties of conflicting community, group and ethnic loyalties.

Through the later part of the 19th century, however, and into the decade of colonial emancipation in the middle of this century, nationalism seemed to hold sway nearly everywhere, often at the expense of constitutional beliefs: ideas that governments should be restrained in some agreed areas by enforceable laws hard to change. The ruling idea has been rather that the nation should have no restraints put upon it, or its devoted leaders: popular sovereignty should be absolute. Perhaps there are some areas of the world where different peoples are inextricably bound together and, even with the best will and fortune, like the Americans (remember the grim and bloody Civil War) would take several generations to evolve a common national identity, if at all. In these areas some of the reasons why constitutionalist ideas first arose may be recreating themselves. National states are not the only possible form of human government, nor necessarily the best.

16 The politics of international intervention

R W JOHNSON

At the conference which gave birth to this book, academician Apollon Davidson of the Soviet Union's Academy of Sciences apologised to the conference (quite unnecessarily, I thought) for his limited English-speaking ability, placing the blame for this situation on the British because "they never conquered my country". His point was well-taken: what Northern Ireland, Israel and South Africa have in common is a starting point under British colonial rule. The key to understanding the differing nature of international intervention in the three cases is to ask how far each has moved away from this colony-metropole relationship. Our three cases may then be placed along a single continuum of internationalisation, Northern Ireland representing the minimum, Israel the medium and South Africa the maximal position.

Northern Ireland: Minimum internationalisation

Northern Ireland remains within a closed colonial relationship with Britain, so much so that both the Ulster Protestant majority and the British government are able to insist that the relationship is not colonial at all but one occurring within a province which is integrally part of Britain itself. This was, of course, the same contention that France once made over Algeria, but this hardly prevented Algeria from becoming a deep embarrassment to France even within the French community. Algeria became a major international issue at the United Nations and elsewhere and the FLN (National Liberation Front) gained wide international sympathy, aid and, ultimately, recognition as an alternative government.

None of these things has happened with Northern Ireland, which is viewed by the international community very much as Britain would wish, that is as essentially a British domestic problem, comparable in principle to Spain's problem with Basque extremism or France's problem with Corsican nationalists. Britain has taken pains to prevent the Northern Irish issue from becoming internationalised Algeria-fashion, but the key reason for British success in this regard has lain in the passive attitude of the Republic of Ireland, for all that the Irish constitution specifically makes a claim for the inclusion of Northern Ireland within a single Irish nation-state.

To see how true this is one has only to imagine how intolerable Britain's position would quickly become if Ireland launched a call for a united Irish Republic: apart from the added difficulties this would produce for Britain in the North, Ireland's legitimate rights in the matter would gain large and rapid recognition amongst the international community. Almost certainly such a call would soon receive the endorsement of the Afro-Asian nations, the Non-Aligned Summit, the Organisation of African Unity and thus majorities within the UN Assembly and even within the Commonwealth. Without doubt the call would also be powerfully taken up by the communities of the Irish diaspora in the United States, Australia and elsewhere. Dublin could, in a word, ignite a virtual forest fire of international support if it so wished — and it is difficult to believe that Britain would have the resolve to stand out against such a sweeping coalition for at all long, particularly given the fact that the British army's continuing presence in the North is hardly popular with the British electorate.

That is, the Northern Ireland problem has not been internationalised, but this could occur quite easily if Dublin so wished, and that internationalisation would have a decisive effect. But Dublin does not so wish. Indeed, when, in 1940, Churchill offered to hand Northern Ireland over to Eire in return for an Irish declaration of war on Germany, President De Valera not merely refused the offer but kept the fact of it secret thereafter. And yet what could have been easier than to publicise the offer once the war was over, and what could have done more to de-legitimate Britain's claim to sovereignty over Northern Ireland than to emphasise that, in the crunch, London had treated the province as a commodity for trade? Yet news of the offer emerged in the end as a result of the official release of British, not Irish archives. That is, in the last analysis, Ireland is keener than Britain that Northern Ireland should remain within the United Kingdom.

Despite this, Britain remains sensitive to the possibility of internationalisation. In particular, Britain has devoted a great deal of effort to attempting to limit the support afforded the Irish Republican Army by the Irish-American community, and to prevent the Irish question from gaining "respectable" support at a formal political level in the US. Generally speaking, this has been successful: the pro-IRA support group, Noraid, has been kept confined to fringe movement status, despite the existence of a far broader current of sympathy and concern over Northern Ireland within the Irish-American community. The coincidence, in the early 1980s, of Irish-Americans in simultaneous occupation of the White House (Ronald Reagan), the House Speakership (Tip O'Neill), the putative leadership of the congressional opposition (Ted Kennedy), and the governorship of New York (Hugh Carey) opened up the dread possibility of a competitive political auction over the Irish question developing within the US. There is no doubt that American pressure was strongly and continuously felt in London in that period, to the extent that the Anglo-Irish Agreement may be seen as its partial result, but, crucially, London was able to keep the issue contained within private and diplomatic channels so that it did not burst forth into public forums of American or international debate.

Similarly, repeated bad publicity over internment, methods of interrogation of IRA suspects, allegations of the use of Special Air Services "death squads" against IRA activists, and of biased or brutal British army behaviour *vis à vis* the Northern Ireland civilian population, has created considerable sensitivity in London at the possibility that the Northern Irish issue might attract international humanitarian or judicial concern. Almost certainly, this implicit pressure has affected British army and policing methods in Northern Ireland: repetitions of "Bloody Sunday" have been avoided, there are now few of the complaints about the indiscriminate use of CS gas seen in earlier years, Sinn Fein is allowed to operate legally, and the IRA is allowed to parade undisturbed in uniform at Republican funerals. This softly-softly approach seems to have worked: despite a number of questionable practices (and despite even a major scandal such as the public murder of IRA activists by an SAS squad in Gibraltar in March 1988), Britain has thus far succeeded in heading off internationalisation on this front too.

The obverse side of Britain's success in preventing the internationalisation of the Northern Irish issue has been the continuing isolation of the IRA. Libya remains the sole state to have recognised the IRA or to supply it with arms (one illicit arms deal with South Africa aside), and while reports circulate continuously about contact between the IRA and other terrorist and "liberation" movements, even some of those contacts may be propaganda fictions. The result is that the IRA, unlike many "liberation" movements, has to buy or steal much of its own weaponry.

The IRA's isolation leaves it relatively unfettered in its tactics. Within Northern Ireland itself the IRA seems to place very few limits indeed on the sort of violence it is willing to carry out — its campaign there bears the hallmarks of communal violence *à l'outrance* rather than a political campaign aimed at eliciting or maintaining public sympathies either in the province or internationally. The targets for such terrorism are not merely the institutions of Protestant supremacy in the North but the whole structure of British rule. At one stage the IRA clearly felt that this rendered legitimate not only attacks on the British military, police and civil administration of Northern Ireland but indiscriminate terrorism against civilians in Britain itself —leading to such atrocities as the Birmingham pub bombing of 1973. Since then such tactics have been abandoned — suggesting that the IRA still places some hopes on influencing British public opinion. Even so, symbolic civilian targets remain on the IRA's list, as the assassinations of Sir Richard Sykes and Airey Neave MP (March 1979) and Earl Mountbatten (August 1979) and the Brighton (Tory conference) bombing (October 1984) all show. Moreover, the IRA has felt free not only to attack military targets throughout the UK but to risk foreign displeasure by carrying out such attacks in Germany and Gibraltar. While the movement clearly still hopes to win foreign sympathy if it can, its behaviour suggests that it sees little immediate hope of breaking out of its isolation. Its frustrations are those of any movement successfully trapped within the confines of a closed colonial system, but it must also suffer the added gall that Ireland may be not only the world's first colony but perhaps also its last.

Israel: Semi-colonial internationalisation

Under the Palestine Mandate what was to become Israel experienced something akin to the closed colonial relationship with Britain that Northern Ireland still enjoys (if that is, indeed, the appropriate term). Indeed, there is a striking parallel between the cases: Britain was wont to argue that it had to stay in Palestine simply in order to hold the ring between warring Jews and Arabs, just as nowadays London argues that it has to stay in Northern Ireland to hold the ring between warring Catholics and Protestants. In that sense, Israel merely represents one of Northern Ireland's possible futures following a British withdrawal.

It could be argued that incipient internationalisation of the Palestine problem provided one of the key pressures forcing British withdrawal. But Israeli independence brought with it a whole new international dimension, as independence inevitably must, if only because other states around the world then have to decide whether to recognise the new state, whether to give it aid, trade with it, arm it, make alliances with it or against it, and so forth. Moreover, independence itself represents a symbolic fact with a powerful new ability to elicit responses right across the international community. An independent Israel had a hugely greater impact upon the Jewish communities scattered around the world than a still-colonial state could ever have had. Moreover, the sheer achievement and finality of that independence was both a proof of the dream among Jews and, among Arabs, a dramatic statement about the permanence of the wrong done to the Palestinians.

Thereafter, the recurrent Arab-Israeli wars had the effect both of widening and intensifying the internationalisation of the Palestine/Israel issue. The 1967–74 period probably saw the highpoint of internationalisation to date, first with the widespread breaking of Third World diplomatic links with Israel in the wake of the 1967 war, and then with the Arab oil boycott of 1973–74 and the consequent change in the international balance induced by the Organisation of Petroleum Exporting Countries. During this period it actually seemed possible for a while that the Arab bloc would succeed in polarising the international community over the Palestine/Israel issue to the same degree that it had been polarised by the Hallstein Doctrine over the question of recognition of East or West Germany, or even as it was polarised by the Cold War: in the extreme case, Israel might even have become as much of an international outcast as South Africa. But the high tide of polarisation receded, first with the Egyptian recognition of Israel in March 1979 (and the gradual failure of Arab rejectionism thereafter), then with the slow collapse of Opec power, and increasingly through the emergence of other Middle Eastern crises in Iran, Iraq and Lebanon.

A key index of the internationalisation of the Palestine/Israeli issue has, of course, been the extent to which the Palestine Liberation Organisation (PLO) has gained recognition around the world. The Palestinian cause made sweeping diplomatic gains in the wake of the 1967 war, despite the fact that the crushing Arab defeat in that war had encouraged the PLO (and other Palestinian groups) to broaden its armed struggle into a wider terrorism. The fact that terrorist outrages (the Black September massacre of Israeli athletes at the 1972 Olympic Games, for example) did not appear to hinder the Palestinian cause from scoring major international gains (such as the oil boycott of 1973–74), may have encouraged the PLO to believe that it could afford to pursue diplomatic and terrorist tactics simultaneously. Certainly, the following decade saw the zenith of Palestinian terrorism — much of it the work of rival Palestinian groups, seeking to outflank the PLO in militancy — which showed an almost cavalier contempt for the diplomatic harm it might do. For Palestinian terrorist groups virtually discarded the Israeli military as targets (they were too difficult, anyway), shifting their attention first to Israeli civilian targets —attacking school buses, hijacking El Al flights and hitting Israeli targets in third countries — and then, when these too became too difficult, attacking Israel's patrons — America, Britain and France. This latter phase saw a series of terrorist attacks such as the Achille Lauro hijacking in which ordinary American civilians (especially American Jews) were treated as legitimate targets. That is, PLO tactics too showed a mirror-image internationalisation in this period.

Despite this high degree of internationalisation, Israel itself has remained too small, weak and threatened a state to be able to escape a quasi-colonial relationship of some kind. For some time after Israeli independence, Britain remained its dominant metropole (and even today Britain is Israel's second largest trading partner), but by the 1950s the French too had assumed a major role, consorting with the Israelis in the 1956 war and acting as Israel's chief arms supplier. However, the 1956 Suez war and the resulting British political debacle effectively removed Britain as a potential metropolitan sponsor: after 1956 it was clear that no British government would be willing to put itself at risk over an Israeli alliance. The 1967 war saw a similar collapse of the Franco-Israeli relationship, leaving the US as the only power willing to act as a metropolitan patron to Israel.

Israel's position thus represents a mixed case. Her position has been thoroughly internationalised. The various European powers all remain major potential players in the region — the 1980s have seen Italian, French and British military forces deployed in Lebanon and the Arabian Gulf. These countries remain of major economic importance to Israel and could, under certain circumstances, be powerful patrons. None of them are, however, keen to play such a role and Israel does not in fact rely upon them. Instead, Israel has a quasi-colonial relationship with the US. The US is the country's chief aid giver, biggest trader, arms supplier and diplomatic protector. It is almost impossible to over-estimate the significance of the relationship to Israel — the US dollar has, in effect, become a second Israeli currency and the US could bring intolerable pressure to bear on Israel merely by threatening to change its tax law on charitable donations. But America is bound to Israel too: the power of the Jewish lobby is such that the adoption of a less friendly

US policy towards Israel is, currently at least, politically inconceivable. In the 1980s the Reagan administration's support for Israel never wavered despite Israel's controversial role in Lebanon and on the West Bank, the discovery of an Israeli spy network operating in the US[1], and the fact that American Jews were the only group other than blacks to cast a majority vote against Reagan, even in his 1984 landslide.

It would, of course, be incorrect to characterise Israel as merely an American colony or puppet state. Israel inevitably retains a freedom of action and manoeuvre that no mere colony can ever have, and the US has not infrequently failed to see eye to eye with Tel Aviv. But the relationship is quasi-colonial, not just because of Israel's overwhelming economic and military dependence on the US, but because Israel's whole place in the world depends, in the last analysis, upon American decision. Thus, for example, Israel can never become as isolated and vulnerable as South Africa while the US-Israeli relationship remains as it is. The US is simply too big, too powerful, its influence too globally pervasive for that. Any state that enjoys the degree of American friendship and diplomatic protection that Israel does, is bound to find acceptance with a large section of the international community. To see the importance of this one has only to imagine what Israel's position might be without American friendship and protection. How would Israel have got through the 1970s in the face of the far more populous, oil-rich and widely influential Arab states if it had not enjoyed American friendship? Which other Western states would have braved Arab wrath to be Israel's friend then, if America had stood aside? And how long could Israel hold out if it became truly isolated, if it became the target of boycotts, sanctions, disinvestment and credit-denial in the way South Africa has?

Thus America is undeniably Israel's metropole. The whole identity and continuity of the Israeli state rests upon American decision. Israel's security rests, in the last analysis, on the fact that the US is unwilling to see Israel ever lose a war to its Arab neighbours. By the same token, all discussions of the Arab-Israeli problem are overhung by the hypothetical fact that the power of decision could be taken right out of Israel's hands. The US can force a "solution" of some sort on Israel if it really wants to (a West Bank state, negotiation with the PLO, even the renegotiation of the exclusive character of the Israeli state). To be sure, the power of the Jewish lobby in Washington means that a high political price must be paid by a US administration that exerts pressure on Israel, but Israel must live with the fact that US policy in this regard is also sensitive to tempting opportunities and pressures (for example from

Gorbachev's USSR) over which Israel has no control.

South Africa: Maximal internationalisation

South Africa represents a case of complete internationalisation because there is no country which is able or willing to act as its metropole. This is evident at every level. Unlike Israel, South Africa has no great power willing to act as its privileged arms supplier, its primary source of loans and aid, its protector in international arenas and forums, even its last-resort human sanctuary (for in practice Israelis worried by the Arab threat can migrate without difficulty to the US). Moreover, whatever their criticisms of Israeli policy, all the Western states — and probably a majority among all countries — view the existence of some sort of Jewish state in the Middle East as legitimate, as likely to continue into the future, and as having a fundamental theoretical defensibility. Indeed, with both Egypt (since March 1979) and Yasser Arafat (since September 1988) now recognising Israel's right to exist, it is possible to argue that the Israeli state even enjoys a growing international legitimacy. South Africa's position is quite opposite: no single state within the international community is willing to argue for the legitimacy of apartheid or white domination, which in turn means that many states deny the legitimacy of the South African state itself.

South Africa is, moreover, seen as a symbolic issue, a final great case of white oppression of black peoples. There is a sense in which Pretoria is being made to pay for white guilt and black rage about the history of slavery in the Americas, the imposition of European colonialism not just on Africa but anywhere, discriminatory racial practices against blacks in Western countries, and so on. Pretoria argues, quite rightly, that there are many cases of racial discrimination elsewhere in the world which do not receive the same attention: East Africa's treatment of its Asians, Chinese racism against African students, gross instances of tribal discrimination throughout Africa, and so on. Perhaps the most remarkable juxtaposition was to be found at the October 1989 Commonwealth conference in Kuala Lumpur. The Malaysian government hosting the conference happily joined in the general denunciation of South African racism although it has itself given racial discrimination the force of law at home. Not only does Malaysian law formally insist that *"Bumiputras"* (sons of the soil, that is, Malays) must be allowed to buy land and property at lower prices than are offered to other races, but the Malaysian statute book includes exactly the sort of job reservation legislation which even Pretoria has now abolished.

Pretoria's tirades against such hypocrisy miss the point. The oppression of blacks by other blacks simply lacks the universal symbolism of white oppression of blacks. Chinese unpleasantness to Africans, Malay discrimination against Chinese, or East African hostility to Asians (let alone the hostilities between Irish Protestants and Catholics) appear to most people outside those countries as mere parochial oddities, perhaps even requiring liberal apologias which attempt to transfer the blame for such behaviour onto the Malay/ Chinese/East African/Irish experience of imperialism. But every person of colour almost anywhere can identify with black experience of white oppression, just as whites almost everywhere can feel guilt about it. The Arab-Israeli confrontation owes some of the international reverberation that it has to the fact that it is often taken to be a partial exemplar of white/black opposition. But South Africa is the full-blown case. In this sense too, South Africa represents a maximal degree of internationalisation. (There is a problem here: the anti-apartheid movement and much of the wider world feels comfortable with the stereotypical black versus white picture of South Africa. It has a historical and emotional "fit". Any notion that changes within South Africa are blurring that stereotype is strongly resisted, for such an acknowledgement threatens to require a new, more complex appreciation — in which some existing moral certainties may be lost.)

Maximal internationalisation is visible in other ways too. There is probably no state in the world which is not on record with its "South Africa policy", usually articulated with considerable zeal. At just about every kind of international forum measures against South Africa (including its expulsion from international bodies) have been at least moved, and often passed. South Africa's liberation movement, the African National Congress, enjoys correspondingly wide international links. Indeed, for some time now it has had representation in more foreign capitals than has Pretoria itself. And this partial exclusion from the international community has led to a strange role reversal. Unlike its Palestinian analogues, the ANC has carefully restricted its extra-legal activities to South Africa itself: even South African airliners and the government's embassies abroad have not been targeted. But the South African goverrment has not shrunk from all manner of illegal activities abroad, including espionage, illegal arms trading (including a deal with the IRA), piracy and assassination. One reason for this is that, as a pariah state already, South Africa has little to lose and, unlike Israel, has no metropolitan state to offend.

The internationalisation of the South African situation is also strikingly visible in the degree of international political involvement there. The ambassadors of the major Western powers quite openly intervene in South African domestic politics, urging the government to allow this or that demonstration, attending this or that political funeral, giving money to opposition causes and so forth. Major companies like Shell place overtly political advertisements in the radical press[2], urging fundamental change upon the government. And a whole host of Western church and charitable groups funnel money to all manner of anti-apartheid groups and activities within South Africa, incidentally producing a class of anti-apartheid political entrepreneurs who become patronage barons through their control over access to such funds. One ironic result of this is that some of those most loudly advocating sanctions and disinvestment from South Africa are themselves the beneficiaries of major inflows of foreign funds. In an age when the collapse of the rand has made international travel an expensive luxury for most South Africans, many of the country's leading anti-apartheid activists have become globe-trotters on a large scale.

Solution by withdrawal

One could multiply the instances of the extraordinary internationalisation of the South African case which marks it out from all others, but the criterion that matters most is simply that the country has no metropole and is therefore beyond the reach of an imposed "decolonisation". There is thus no "solution" available in the South African case of the sort that is available even in Israel and Northern Ireland.

For the fact is that in the last analysis the US could force a "solution" on Israel, just as Britain could force one on Northern Ireland. This could be done in the classic decolonisation style, that is, by metropolitan withdrawal. The withdrawal of metropolitan economic assistance would be sufficient to threaten the collapse of the Israeli and Ulster states, and force Israelis on the one hand, Ulster Protestants on the other, to reach some form of accommodation with their neighbours. The fact that most Israelis and Ulstermen would furiously resist such a "solution", and indeed, would deny that it was a solution at all, does not alter the brute fact of the metropole's ability to transform the local situation.

The chief objection to metropolitan withdrawal is, of course, that only the continuing involvement of the metropole guarantees the fragile balance. Withdrawal, it is argued, would not only imply a betrayal of the Israelis/Ulstermen but would in any case produce no new equilibrium, merely a descent into a nightmare of uncontrollable violence. The paradigm case is that of Algeria, where French withdrawal saw the escalation of OAS (Organisation de l'Armée Sécrete) violence, together with answering reprisals from both the French authorities and the FLN. In the space of a year the OAS killed 2360 and wounded 5418. In its last six months alone the movement claimed three times as many civilian casualties as the FLN had in six years.[3] In addition, the OAS destroyed literally thousands of buildings, including schools, laboratories, hospitals, the BP plant at Oran, municipal buildings and the University of Algiers library.

The French army was finally moved to respond on a parallel scale with the attack on the suburb of Bab-el-Oued (an OAS stronghold) by strafing fighter-planes and 20 000 troops armed with tanks, rockets and heavy machine guns.[4] FLN reprisals were at first comparatively restrained but escalated over time — and when the FLN marched triumphantly into Oran a wave of Muslim militants ran through the half-deserted European quarters cutting the throats of such men, women and children as they could find.[5] In the final convulsion close to one and a half million settlers and *harkis* (loyalist Muslims) fled Algeria in the space of a few months, while the FLN exacted a terrible vengeance on the thousands of *harkis* who remained behind. Prior to this Algeria had run almost the full gamut of colonial crisis, with repeated attempts at UDI by the army and the settlers, endless abortive schemes for "reform", and repeated threats of partition, with the question of the Sahara remaining unresolved almost to the end.

The Algerian nightmare was real enough — and, of course, the inter-communal violence which accompanied Indian independence occasioned even greater casualties and population removals — but it is worth pointing out that similar outcomes have been threatened far more often than they have actually occurred. In the era of colonial independence many countries heard dire threats of a "scorched earth policy" by one or other of the groups struggling for supremacy, but Algeria was extremely unusual in seeing such a policy actually carried out for a time.[6] Similarly, inter-communal tensions within colonial states have also led to frequent dire warnings of the inevitability of partition in the event of metropolitan withdrawal, but the cases of India/Pakistan and Ireland/Ulster are even more unusual. (Interestingly, De Gaulle considered but explicitly rejected the option of partition in Algeria on the grounds that it would merely repeat the foolishness of the English in creating a Northern Ireland problem to dog them down the years.[7])

Finally, it is worth pointing out that, terrible as the Algerian nightmare was, it was also brief. Despite the OAS's commitment to fight forever for *Algerie Française* and never to tolerate those who truckled to negotiate with the FLN, the OAS itself negotiated a cease-fire with the FLN three weeks before independence.[8]

Attempts by their metropoles to force solutions on Israel and Northern Ireland would be met with all the same warnings of bloodbaths, inter-communal violence, UDIs, fresh partitions, scorched earth policies and endless strife. The Algerian example shows that these possibilities are, at once, real, rare and brief: they need not deter the determined metropole, whose weapon is always the same: withdrawal. The same dire warnings are heard in the South African case but here there is no metropole to force such a solution. In this case, however, we have, uniquely, seen the formation instead of a "collective metropole".

The collective metropole

It is only over the past decade that in South Africa's case we have seen the emergence of a "collective metropole", that is, a loose assemblage of Western powers attempting to use their leverage to enforce a "de-colonisation" of South Africa. The assemblage is so loose that it is difficult to specify membership precisely — really one is talking of the entire OECD (Organisation of Economic Co-operation and Development) bloc, that is, all the West European, Scandinavian and North American states plus Japan, Australia and New Zealand. Despite the width of this group there has been an increasing concertation of policies towards South Africa amongst all its members. Thus, for example, the question of the UN arms embargo against South Africa was treated separately by each OECD state, but gradually more and more, and then all, fell in line with it. Amongst the largest constituent group, the West European bloc, there are now regular and formal attempts to harmonise policy towards South Africa, while Japan, until recently something of a free-rider on the South African issue, has been pulled into line by American pressure.[9]

There is, moreover, a discernible core group to this OECD "collective metropole" — the UK, US, West Germany and France — which is able to give direct diplomatic leadership and expression to the OECD's generalised pressure. This core first saw action in the shape of the Western Contact Group in the Namibian negotiations, in initiatives such as that of the Eminent Persons' Group, in the clearing of PW Botha's abortive "Rubicon" speech with the major Western embassies, and in a host of diffuse but continuous diplomatic contacts and pressures. Within this core, I have argued elsewhere,[10] the pivotal state is the UK. It alone is a permanent member of the UN Security Council, of the European Economic Community, of the Group of Seven, and the Commonwealth. That is, while the UK can no longer control South Africa's international environment (as it once did when it was South Africa's sole metropole), it can to some extent regulate changes in that environment. Thus, for example, there cannot be generalised sanctions against South Africa while the UK refuses to go along with such a move. Inevitably, the UK's special hinge-position has meant that the British government has come under exceptional international pressure over the South African issue in recent years.

This constellation of forces has resulted in some extremely complex lobbying processes. Indeed, given that the whole international community is involved in this issue, the action and reaction it produces is almost unlimited. Until well into the 1960s the South African government had its solid coterie of established friends and the ANC, South Africa's principal "liberation movement", found its support confined principally to the communist bloc, radical African states and liberal and left-wing groups in the West. This created a sort of stand-off, for while ANC support was extensive it was confined to circles with minimal leverage over Pretoria. Pretoria was thus able largely to ignore foreign pressure in this period, denouncing criticism as an *ultra vires* attempt to interfere in its domestic affairs, and almost making a virtue of thumbing its nose at international opinion.

Gradually, the balance changed. Essentially, the ANC played to its strength among the Afro-Asian and communist blocs and these forces then exerted pressure upon the OECD states. The first sign that this was working was the UN ban on arms exports to South Africa. One by one, under strong Afro-Asian pressure, the OECD states fell into line with the ban. Significantly, when the Heath government came to power in Britain in 1970 with a clear commitment to overturn the ban and resume arms exports to South Africa, African pressure exercised via the Commonwealth was so fierce that Heath had to back down.[11] Of all the countries with sophisticated armaments industries, the French were thus left isolated as the only state still willing to sell arms to South Africa. France joined the ban in 1977 after French Foreign Minister Louis de Guiringaud found his tour of East Africa disrupted by protests over French policy towards South Africa. Moreover, President Nyerere announced publicly that he could hardly rein in such demonstrators since his government entirely shared their feelings about French arms sales to South Africa. The possibility clearly loomed that such protests would become generalised elsewhere in Africa —and beyond, threatening the dignity, order and good name of French diplomacy. Thus here again the ANC tactic of pressuring the Western states via African intermediaries did ultimately work.[12]

Playing to the gallery

As the government and the ANC move into the era of negotiations, this gallery-playing has reached a new intensity, with each side attempting to wrong-foot its opponent by showing that it is the more "reasonable". The ANC has now achieved official contact with most OECD governments but the primary galleries to which it plays are still the Frontline states (FLS), the OAU, and the Afro-Asian and communist blocs. The process is most clearly seen in the run up to Commonwealth conferences. The ANC puts its position to the FLS, who then take the lead in agreeing a line which becomes that of the OAU. This line is then pressed on Australia, Canada and New Zealand and at the conference itself all parties exert pressure on Britain.

The run-up to the 1989 (Kuala Lumpur) Commonwealth conference illustrates how complex the gallery-playing has now become. Well in advance of the conference there was extensive contact between the Congress of South African Trade Unions (Cosatu) and the Australian Foreign Office and leaders of the United Democratic Front/Mass Democratic Movement (UDF/MDM) were received in Washington by President Bush and in London by the Foreign Office. In August 1989 the ANC produced its document on "Guidelines to the Process of Negotiation" — its timing doubtless influenced, perhaps even determined by the approaching (October) conference. This draft was first cleared with the Frontline states, then rushed to Addis Ababa where it was adopted by the OAU special committee on Southern Africa, which felt free to speak for "the people of Africa, singly, collectively and acting through the OAU". The draft was then whisked on to Belgrade where it received the benediction of the Non-Aligned Summit conference. The draft was then taken to Kuala Lumpur both by the FLS and by the ANC (which maintained a presence, so to speak, in the corridors of the conference), and an attempt was made to have the draft endorsed in turn by the Commonwealth itself.

While all this was going on Pretoria had been busily playing to some of the same galleries — a process really begun the year before by President PW Botha, who had visited Europe as well as a number of neighbouring African states, but rapidly continued by his successor, President FW De Klerk, who paid visits to Mrs Thatcher in London and to President Kaunda in Zambia. In addition, a good deal of diplomacy was carried on less visibly by both sides, as also Pretoria's quiet negotiation with international bankers in Switzerland of a roll-over of South Africa's debt. News of this debt renegotiation broke during the conference and, together with Mrs Thatcher's resistance to the imposition of further economic sanctions against South Africa, it meant that Pretoria had won this particular round. For, at the end of the day, all the sound and fury of the FLS, the OAU and the Non-Aligned Summit only fully serve their purpose if they have the effect of altering the behaviour of South Africa's "collective metropole". On this occasion they did not — but there will be other rounds to fight.

Two other aspects of this gallery-playing are worthy of note. The ANC's "Guidelines" document was preceded by a statement of principles on which a new South African constitution was to be based, and included such items as a guarantee of all "human rights, freedoms and civil liberties", with an entrenched bill of rights and an independent judiciary in a democratic state. This document was happily endorsed by the Frontline states, the OAU and the Non-Aligned Summit despite the fact that a clear majority of members of each of those forums do not themselves favour such constitutional principles in the governance of their own countries — a fact which serves to emphasise the point that they were never intended to be the ultimate consumers of this document, which was pretty clearly targeted on the OECD states to whom some sign of liberal constitutionalism is important. That is, the international galleries and constituencies to which both sides make appeal are actually *shaping the substantive political content* of the struggle. The South African government is subject to the same phenomenon, as was evidenced by the way that PW Botha's abortive Rubicon speech was vetted by the embassies of the "collective metropole".

The ANC's "Guidelines" document also contains a demand worthy of particular note: that while a new democratic constitution is being drawn up the country should be governed by an interim government made up of "the South African regime" and "the liberation movement". In effect this would mean that the ANC be given a large measure of power prior to any consultation of the electorate. One of the uses of the draft document is that this pre-emptive bid for power has now received wide international endorsement and is on the table for discussion in all other forums where the South African issue arises. Thus although the

Commonwealth conference did not adopt the "Guidelines" document, even British Foreign Office officials at the conference suggested that the document might provide one of the key references for the drafting of the Commonwealth's own position. The move well illustrates the way in which gallery-playing can be used to place items on the agenda, so that all future lobbying will involve inclusive pressure for those items.

Internal decolonisation?

The "Guidelines" also provided that the two contending parties (Pretoria and the ANC) "shall define and agree on the role to be played by the international community in ensuring a successful transition to a democratic order". This raises the key question of what role the "international community" *can* actually play.

Elsewhere in Africa white-ruled regimes have ultimately been forced to bow to majority rule by two factors: the growing inability of the settler community to protect itself from its local "liberation movement", and the metropole's threatened or actual withdrawal. In South Africa's case the ANC's armed wing, Umkhonto we Sizwe (MK), has not to date presented a serious threat to the South African state and it seems unlikely that it ever will. But the withdrawal of the (collective) metropole has been going on for some time, with the progressive removal of cultural and sporting contacts, military treaty arrangements, air landing rights, arms deliveries, trade, investment and credit. This withdrawal has already exerted great pressure on Pretoria and the threat of worse to come has pushed the government all the way into a commitment to negotiations with the extra-parliamentary opposition, including the ANC.

The fact that South Africa conforms to only one side of the decolonisation model means, however, that there is a certain artificiality to this negotiation phase. Neither the government nor the ANC is psychologically ready for proper negotiations and the internal balance of power hardly necessitates negotiation. Strong and continuing presence from the collective metropole may well be necessary to achieve progress towards a settlement, but the metropole's role on the ground is open to question.

Mediation

Pretoria is adamant that it never wishes to see a "Lancaster House" conference of the type that brought independence to Zimbabwe. In practice Britain under Mrs Thatcher has already become a *de facto* mediator: Mrs Thatcher saw President De Klerk soon after he took over the National Party leadership and hectored and lectured him in typical fashion before going off to Kuala Lumpur to be lectured and hectored by the Frontline states and the rest of the Commonwealth. Thus critical pressures and communications pass back and forth through the pivot of the British premier, which means that other countries of the collective metropole see hers as a key reference position. It was notable that Mr De Klerk took pains to communicate advance news of the release of Walter Sisulu and the other Rivonia detainees to Mrs Thatcher in October 1989. But this crude mediation might well not survive Mrs Thatcher's departure from the premiership and neither Pretoria nor the ANC wants her role to grow.

Probably no single country will have the degree of bona fides and neutrality necessary to be freely acceptable both to Pretoria and the liberation movement(s). But this leaves open the possibility of "forced mediation" along the lines of the Namibian model. In that case, the countries of the collective metropole (particularly the US) collaborated with the USSR to force South Africa to settle, the Frontline states to go along with the settlement, and Swapo to comply with the agreement even though it had played no part in reaching it. In the South African case it seems unlikely that the ANC could be treated in similar fashion or, if it was, that the FLS would accept any resulting agreement. But combined US-Soviet pressure could probably force the ANC, however unwillingly, to accept mediation. Even to state the problem thus highlights the fact that the one common interest of Pretoria and the ANC would probably be to avoid this sort of mediation.

Peacekeeping

A critical part of the Namibian package was the peacekeeping role played by the UN Transition Assistance Group, just as the British army had played a similar role in Zimbabwe. This, crucially, is unlikely in the South African case.

It has to be remembered that the final concession of majority rule in South Africa could, at worst, trigger a magnified form of the type of response seen in Algeria. Right-wing whites could well conduct a campaign along OAS lines (and there are five times as many whites in South Africa as there were in Algeria), which in turn would trigger retaliation from the government, and from the ANC. Competing black groups such as Inkatha or the PAC could well react with similar intensity if they felt that the settlement deprived them of a reasonable chance of accommodation. In such an environment large-scale white and Indian emigration would occur, together with capital flight and economic collapse, triggering more routine forms of insecurity as well.

It is not at all difficult, in a word, to conjure up a picture of virtual *Gotterdammerung*. One only has to study the speed with which Israel, the US, Britain, France, and the combined Arab states have all abandoned the idea of a peacekeeping role in the comparatively tiny case of Lebanon to realise that it is in the highest degree unlikely that the UN or any other outside powers would be at all keen to send a peacekeeping force to South Africa, even if it were welcome — which it would not be.

Election certification

This is a role which the collective metropole could usefully play, rather as the Commonwealth Observer Mission did in Zimbabwe or Untag did in Namibia. The problem is that there would be no peacekeeping force back-up to such a mission and anyway both the ANC and Pretoria would doubtless realise that there is a certain fraudulence to such missions.

In practice all such elections have to be certified as acceptably free because it is unthinkable to begin the whole process again from scratch: the world community wants to see a solution; the metropole is unwilling to undertake the risks and expense all over again; and to invalidate the election is to make a mortal enemy of one party or another.

Such considerations seem to have led Lord Soames to accept the election result in Zimbabwe despite gross irregularities,[13] while the UN Representative, Mr Ahtisaari, seems to have taken a fairly broad-minded view of somewhat lesser irregularities in Namibia. So while international election observers may well get the contending parties to behave better than they otherwise might, it must be realised that they are merely the icing on the cake.

Guarantee of the settlement

The metropole may, as a means of making a settlement more acceptable, offer to guarantee that it will not tolerate the breach of certain provisions. Thus, in the case of Zimbabwe, Britain insisted on minority representation for whites in parliament for a period, together with full compensation for white farmers who were bought out.

The implied back-up threat to such guarantees is the use of sanctions against the successor regime if it does not comply. This was not, in fact, very plausible even in Zimbabwe's case and would be fairly laughable in the South African case. It is unthinkable to imagine an international campaign of sanctions being led against an ANC-ruled South Africa, whatever the ANC did: even if ANC rule was followed by large-scale massacres of whites it is probable that many within the international community would feel that the whites had brought such an outcome upon themselves. But whatever view was taken, a large-scale intervention to keep the peace would remain as unlikely as before.

This is not to say that a formal "guarantee" by the countries of the collective metropole of a negotiated solution would be wholly worthless — no South African government would want to be seen to renege in the face of such an assemblage of might — but such a guarantee would be largely bluff. It would certainly be useful if any settlement did carry such a guarantee but it is unrealistic to imagine that Pretoria would pay a very high price to obtain it.

What all this suggests, in sum, is that while the collective metropole is in a position to force Pretoria along the road to negotiations, it is not effectively in a position to play a strong part in any of the other "normal" metropolitan roles. It may be able to effect a certain degree of mediation, to certificate elections and to give a paper guarantee of a settlement, but its leverage (and thus control) over the situation is likely to remain at arm's length throughout. Its critical weakness is its unwillingness, perhaps even its inability, to mount a sufficiently strong force on the ground to hold the ring during the period of transition to majority rule.

Conclusion

To discuss such scenarios is to bring home the fact that South Africa will, uniquely, be responsible for its own decolonisation. There will never be a point when only a metropolitan army stands between South Africa's whites and the vengeful army of a triumphant liberation movement. One reason is that no such metropolitan army will be deployed. Another is that Umkhonto has never remotely threatened to move away from low-level guerrilla tactics. But above all, the SA Defence Force and SA Police are bound to remain by far the most powerful armed forces in South Africa throughout the transition to majority rule. That is, South Africa must hold the ring for itself while this necessary and inevitable transition takes place. This conclusion has several major corollaries.

First, it is essential that, as far as possible, the whole apparatus of law and order — including not just the police and army but the intelligence services and the judiciary — becomes more neutral and is perceived as such. Happily, this no longer appears quite such an outlandish hope as was once the case. Some members of the judiciary (particularly in Natal) have been displaying a more enlightened and independent stance for some time now, including a reluctance in some cases to use the death penalty — a key point, given that most of those executed are black and that such sentences leave a lasting bitterness.

Some policemen have also shown signs of a more independent and professional attitude. SAP Captain Ray Harrald won praise from all sides for his role in bringing about a peace settlement in Shongweni (Natal) between the contending UDF and Inkatha forces in August 1989.[14] The same month saw the Durban police raise a number of eyebrows by the way in which they co-operated with MDM stewards to ensure that a pre-election "beach invasion" remained peaceful — indeed, a police spokesman described the demonstration as "a feather in the cap of the MDM".

The following month saw Lieutenant Gregory Rockman speak out fiercely against the actions of the riot police in the Western Cape and continue his criticism even in the face of disciplinary action against him.[15] There were, too, other straws in the wind, such as the attempts to form a police trade union and the "death squad" revelations of former policeman Dirk Coetzee.[16]

Second, the De Klerk government would be wise to show, by word and deed, that it will not tolerate unprofessional behaviour among the police. Tremendous good would be done by a determined attempt to forbid and root out the use of torture, by the abolition of the death penalty, and by the appointment of more open-minded judges and magistrates. The appointment of some black judges and police chiefs would be an obviously sensible move, particularly if some of the appointments went to individuals identified with the opposition.

Third, the MDM/ANC could contribute greatly to the evolution of a more neutral and professional law and order apparat if it were willing to show appreciation of more enlightened police/judicial behaviour whenever it occurs. The notion that all policemen, all judges and all soldiers are incapable of behaving even-handedly —while perfectly forgiveable now, as a judgement on the past —will ill serve the future. Policemen and judges alike need to feel that professional neutrality on their part will be rewarded by genuine esteem. The same goes for the whole state structure, indeed, the whole of civil society.

Such considerations assume that the model of a liberal state is truly possible in South African conditions, that one really can think in terms of equality before the law, a bill of rights, and pluralist political competition. All that one can say is that perhaps the chief objective of internal decolonisation ought to be to install these conditions as fully as possible and to hold the ring long enough for competitive democratic elections to take place not once but a number of times. There should, in other words, be no single "independence election" but the gradual routinisation of a process of liberal political competition. At the end of the process, no doubt, power would have been transferred from white hands to black ones, but the real success of internal decolonisation would be that no one would ever be quite sure when "independence" took place.

ENDNOTES

1. The Pollard affair, uncovered in 1985. Neither R Deacon: *The Israeli Secret Service*, Hamish Hamilton, 1977, nor D Eisenberg *et al: The Mossad*, Paddington Press, 1978, make any mention of Israeli intelligence activity within the US and, given the close links between Mossad and the CIA, the main surprise was that such activity was even necessary. Pollard was sentenced in April 1986. Only a few months later the Irangate scandal broke, revealing that Israel, at America's behest, had simultaneously been acting as a secret conduit of US missiles to Iran. The speed with which both these affairs were buried was no small testament to the power of Washington's Jewish lobby.

2. Thus Shell takes out regular whole-page adverts in the *Weekly Mail*, for example in *WM* October 17 to November 2, 1989: "FREE THE DEMOCRATIC PROCESS. Shell urges government to: 1. End the State of Emergency 2. Release and unban all political leaders 3. Lift restrictions on democratic organisations 4. Allow and encourage freedom of expression." Such ads not only give Shell a high and explicit political profile but directly subsidise and protect a newspaper which the government has several times banned and threatened.

3. A Horne: *A Savage War of Peace. Algeria 1954-1962*, Macmillan, 1977, p 531.

4. Horne, op. cit., pp524-6, 530. This was pretty much what OAS militant Roger Degueldre meant when he vowed to "do a Budapest", op. cit., p483. See also P Viansson-Ponté: *Historie de la République Gaullienne. Tome I. La Fin d'Une Epoque*, Fayard, 1970, pp448-59.

5. Horne, op. cit., p533.

6. Horne, op. cit., pp528-30.

7. See Viansson-Ponté: *Tome I.* pp263, 337, 325 and also J Lacouture: *De Gaulle. Tome 2. Le Politique*, Seuil, 1985, pp604-24 and *Tome 3. Le Souverain*, Seuil, 1986, pp49-283 *passim*. From the time of Pierre Lagaillarde's attempted coup in Algiers in January 1960, one of the options frequently discussed within the settler population and the army was of "creating a new South Africa". De Gaulle, for his part, entertained the notion of a partition which would leave the Sahara either in French hands or under a Franco-Algerian condominium. The settlers had no concern for the

Sahara — they all lived along the Algiers-Oran coastal strip, not in the desert — but De Gaulle wanted the Sahara for French nuclear testing. During the Evian negotiations with the FLN he repeatedly threatened a partition along the lines the settlers might have preferred, with the army guarding a white coastal enclave from which the Arab population was expelled (see Viansson-Ponté: *Tome I*, p403) — but this was merely an attempt to exercise leverage over the FLN. The FLN took the threat seriously enough to stage a "national day against partition" throughout Algeria on 5 July 1961, the demonstrations leaving 80 dead and 266 wounded (Viansson-Ponté: *Tome I*, p403).

8. Viansson-Ponté: *Tome I*, p531. The OAS, for all its savagery and boasting, did not long outlast Algerian independence. In February 1963 its main military leader, Colonel Antoine Argoud, was arrested and sentenced, causing his co-leader, former French prime minister Georges Bidault, to flee to Brazil. Other OAS activists attempted to flee to Latin America too but many were caught en route. See Viansson-Ponté: *Tome II. Le Temps des Orphélins*, Fayard, 1971, pp96-7.

9. Japan's emergence as South Africa's biggest trading partner in 1987 led to a good deal of adverse comment in the Third World and by the US, causing the Ministry for International Trade and Industry (MITI) in Tokyo to ask major companies to limit their South African trade. The fruits of this were seen in 1988 when Japan fell back to become South Africa's second biggest trading partner — behind West Germany.

10. For further elaboration of this notion of the "collective metropole", see R W Johnson: "How Long *Will* South Africa Survive?", *Die Suid-Afrikaan*, February 1989, and "Back to the Future: Looking back on 'How Long Will South Africa Survive?'" in P Collins (ed): *Re-thinking South Africa*, Harvester-Wheatsheaf, forthcoming 1990.

11. Mr Heath did formally abrogate the arms embargo but ran into a storm of criticism over the issue at the January 1971 Commonwealth conference in Singapore. A joint British-Commonwealth team was set up to investigate the issue but in effect Heath's initiative was dead. Britain did supply some helicopters and spare parts to South Africa, but that was all. The embargo was confirmed by the next Labour government and when the Conservatives returned to power in 1979 no attempt was made to revive the issue.

12. The history of French compliance with the UN arms embargo is instructive. President Giscard d'Estaing first announced compliance with the embargo in August 1975 during a visit to Zaire and following a Franco-African conference at Bangui in March 1975 where the issue was doubtless raised. Giscard had given the go-ahead for the sale of four Agosta submarines to South Africa in June 1975 and made it clear that France would maintain naval deliveries as also existing contracts for air and land weapons. (*Keesings Contemporary Archives 1975*, p27324) This was followed by the announcement on 1 June 1976 that France was supplying 16 Mirage F-1s to South Africa, with 36 more to be built under licence in the Republic. Giscard might have got away with this but for the outbreak of the Soweto Uprising two weeks later. By the time the OAU Council of Ministers and Assembly of Heads of State met in Addis Ababa in July 1976, feelings were running extremely high and France was vociferously condemned — the OAU deputy secretary-general actually going to Paris to denounce French policy publicly. De Guiringaud's ill-fated East African trip followed. The situation thus produced was a grave embarrassment to Giscard, threatening his attempts to set up a North-South conference in Paris and queering the pitch for his attempted resumption of relations with Mali and Guinea, neither of which had been visited by a French president since independence. Giscard ultimately announced full compliance with the embargo in February 1977 while on a visit to Mali and in the following year he was able to visit Guinea too. *Keesings 1976*, p28424; *1977*, p28372.

13. The eight supervisors of the election commission sent to Zimbabwe by Britain found

evidence of intimidation by the Muzorewa forces in Victoria province but far more widespread intimidation by Mugabe's followers. Zanu-PF behaviour was so severe, they reported, that conditions for "fair and free" elections did not exist in five of the eight provinces. In Victoria province Zanla fighters threatened death sentences to anyone not supporting Mugabe and one Zapu candidate was last seen having burning coal stuffed down his throat. Lord Soames repeatedly reprimanded Mugabe over intimidation and Mugabe was desperately worried that the elections might be called off or the result cancelled. See D Smith, C Simpson and I Davies: *Mugabe*, Sphere, 1981, pp180-99. This intimidation was possible largely because several thousand Zanla guerrillas managed to avoid being confined to camps and wandered round the country exerting brutal pressure. It seems likely that Swapo's abortive armed incursion into Namibia on 1 April 1989 was motivated by a wish to gain similar leverage over the Namibian elections.

14. *Natal Mercury*, 30 August 1989.
15. *Weekly Mail*, 29 September to 5 October 1989 and 13 to 19 October 1989.
16. *Weekly Mail*, 24 to 30 November 1989.

17 The elusive search for peace

HERMANN GILIOMEE

An insoluble problem?

This book addresses the question of why the search for peace in South Africa, Northern Ireland and Israel is so elusive. Most of the chapters show that what is at stake is not only competition between ethnically defined groups over scarce socio-economic resources; in each case there has also been a larger national struggle over self-determination, sovereignty and survival. This final chapter draws some conclusions about the similarities and differences in the conflicts before suggesting requirements for peaceful accommodation. While drawing on the individual chapters, the conclusion represents the opinion of the author alone.

The three societies under discussion are all divided societies — the general term for societies in which conflicts take place between groups and not classes, and where the groups are defined ascriptively, whether it be on the basis of race, ethnicity or religion. The basic political unit is the racial or ethnic or religious group, and where they have the chance people continue to vote overwhelmingly for parties representing the respective segments. Political violence, whether vigilante or revolutionary, comes to be seen as representative violence, that is "as representative of the community from which it emanates and for which that community is held accountable".[1] Even intra-communal violence is seen as a product of the wider struggle.

There are features of the conflicts in Northern Ireland, South Africa and Israel which give them a particular character. Firstly, their dominant groups share an important link with the West, or more particularly Europe. These groups believe their history to be one in which "progress" and "civilisation", defined in European terms,[2] were brought to a land the natives were using unproductively. Politically, a provocative dualism developed: for the settlers and their descendants there was democracy as Europe understands it: for the natives and their off-spring colonial subjugation.[3]

Despite this dualism the ruling groups have continued to stress the probity of their actions in terms of Western values. The modern-day descendants of the settlers still believe themselves worthy of and entitled to the privileges and rights they enjoy. In Israel and Northern Ireland these sentiments are strengthened by the fact that the dominant group represents a majority within a democratic system. However, even under the apartheid system the dominant group displays few misgivings. As Gagiano's chapter shows, white students value the current system of rule not only for security reasons but even more because they consider it to be good, honest, free and just.

In the second place, all three dominant groups have appropriated the land not only for material gain but also to underpin their national identities. They are, in other words, homeland peoples. There has been an "obsessive search for rootedness",[4] a continuous attempt to take possession of the homeland, fatherland or *patria*. An extension of this is the dominant group's claim that it has "nowhere else to go". A poet from the Ulster Protestant community has written: "This is my country, nowhere else, and I shall not be outcast upon the world".[5] It is a sentiment that resonates powerfully among Afrikaners, even more so among Jews haunted by the recent history of the Holocaust.

Given this background it is hardly surprising that caution is the watchword of communal leaders. Far from enjoying the latitude of dictators in some divided societies, they can be censured or removed both by their party and by the electorate. They are closely watched, for outcomes in conflicts over national identity are generally seen as of a zero-sum and irreversible kind, allowing little room for risks.

Thirdly, there is something unique to the struggle of the subordinates in the three societies discussed in this book. Within their own countries they are confronted in microcosm by Europe's exploitation of the Third World, and with the entire European tradition of cultural chauvinism which juxtaposes assumptions about its own economic and technocratic superiority with assumptions about Third World "backwardness".[6] More than in other divided societies, these subordinates have been turned into strangers in their own homelands.

The conflicts in all three societies are national, rather than civil rights struggles, although the conflict is articulated in different ways in each particular case. A national struggle implies that it is virtually impossible to settle discrete social problems, such as poor education or public health problems, if the national issue is left unresolved. A study of Northern Ireland from a comparative perspective makes this important point: "To describe a conflict as national is to say that, in disputing which community should have the last word, it has become a conflict about everything because particular issues become difficult to isolate".[7] Oliver Tambo states the position of the African National Congress (ANC) in the mid-1980s unambiguously: "This is not a civil rights struggle at all ... our struggle is basically, essentially, fundamentally a national liberation struggle".[8] Nusseibeh makes the same point about the Palestinian struggle in similar words (see Chapter 7).

Nevertheless, it is important to note the different ways in which the struggle expresses itself in each individual case. Some analysts have made a distinction which identifies the conflict in South Africa as one over race, while those in Israel and Northern Ireland are perceived as being about religion. They argue that the latter conflicts are intractable because religion involves absolute truths. For the sectarian thinker "there is no need to look further than the faith and the consequent depravity of their opponents for explanations of the violence".[9] By contrast the situation in South Africa is regarded as more open to compromise because race as a concept is now generally discredited and because one can bargain over the extent of racial privilege.

This perspective is not particularly illuminating. The important point is that race (or religion) has become fused with ethnic, communal or national feelings. It is the latter dimension which makes claims on the basis of history and contractual obligations. Thus while the Protestants of Ulster do object to a Catholic-dominated state on doctrinal grounds, they are animated even more strongly by their conviction that the United Kingdom has a historic and contractual duty to protect the Union and Loyalist community.[10] Similarly, as O'Malley's chapter shows, Northern Ireland Catholics do not object so much to the Protestants' religion as to their politics which gives them the final say.

Israel's rejection of a Palestinian state has less to do with religious differences than with the security threats the state has experienced in its brief history and its abhorrence of past Palestine Liberation Organisation (PLO) terrorism. This feeds into the "people apart" syndrome of the Jews and a very strong tendency to "go it alone" rather than to depend on alliances with neighbouring states to cope with national security problems.[11]

Finally, while doctrines of race are discredited even in South Africa the historic claims and cohesion of the Afrikaner and the larger white community have not simply withered away.[12] This is due partly to a common position of privilege but equally important is the desire to preserve in South Africa a First World identity which is now being reformulated in a racially inclusive way.[13] A more muted but still significant ideological tendency is Afrikaner ethno-nationalism, which is demonstrated empirically by Gagiano's chapter. While the National Party and Conservative Party have sharply conflicting views about political strategy, they share the common goal of securing, albeit in different ways, the political destiny of the Afrikaners.

Prospects for a settlement

Analyses which highlight the intractable nature of the struggle in the three societies invariably provoke the reaction that this interpretation almost automatically rules out any settlement. Yet history has strange twists and turns. Very few analysts expected the Sadat initiative in 1977 to make peace with Israel or the willingness of Israeli Jews to give up the Sinai peninsula. The world has been equally surprised by South Africa's about-turn in legalising the liberation movements and starting talks with the ANC.

In the literature on conflict resolution a considerable emphasis has recently been placed on the concept of "ripeness". This perspective argues that parties settle for peace when they conclude that the status quo is intolerable and that the costs of persevering with it are higher than the costs of a compromise between the adversaries. It is further suggested that much depends also on whether the chemistry between the leaders on both sides of the political divide is favourable. These insights are useful in their own way and can indeed be used to explain the decision of the South African government and the ANC to negotiate with each other. However, "ripeness" can only be established with hindsight. The remaining part of this chapter will explain the conditions favouring the situation becoming ripe for a settlement. Three variables are crucial:

(a) the involvement of external patrons or guarantors;
(b) the ability of the adversaries to reformulate their respective identities and interests in ways which make accommodation possible;
(c) the capacity of the political systems of the adversaries to absorb pacts and unpopular compromises made by leaders.

The involvement of external patrons

It is important to distinguish between a conflict which is internationalised and one where there is active involvement of external patrons or guarantors. In the chapters by Guelke and Johnson, Northern Ireland is presented as the least internationalised conflict of the three under discussion. To be sure, Northern Ireland is rarely an item for debate in international forums, yet of the three societies discussed in this book, it is by far the most affected by external interference. This profoundly shapes the hopes and fears of the disputants as they consider their political options.

As a rule, power-sharing between groups succeeds where the prospects for external intervention are so remote "that no one except a few maniacs hope for such external support to cancel the power of their internal opponents — or ones who are lucky enough never to have external allies to dangle tempting offers before them".[14] An example is Belgium, where ethnic tensions have tempted neither the Dutch nor the French to intervene on behalf of the Flemish or Walloon segments.

By contrast, the conflict in Northern Ireland has become so intractable precisely because the interference of the British government and the Republic of Ireland has blocked the internal political process. Over the past two decades the Unionists have made very few attempts to court the goodwill of Nationalists. Fearing above all British betrayal of the Unionist cause, they try to shield themselves against it by appearing ever more intransigent. For their part, the Nationalists, both in the Republic and in Ulster, have done precious little to persuade the Unionists of the desirability of a united Ireland. Nationalists have tended to "lean on the British to lean on the Unionists to move towards a united Ireland".[15]

The Anglo-Irish Agreement of 1985 tries to build a settlement on the very slippery ground that makes the conflict so intractable. It institutionalises the patron roles of both Britain and the Republic, assigning to them the function of holding the ring to enable the internal parties to come to an accommodation. However, the Social Democratic and Labour Party (SDLP) still attempts primarily to persuade the British to become pro-Irish unity, in the expectation that once Britain is so persuaded, unity will inexorably follow. If there is any Nationalist attempt to address the Unionists, it is to encourage them to become disillusioned with the British or to warn of the dire consequences for Northern Ireland if the Unionists fail to let themselves be convinced.

But even if the Unionists allowed themselves to accept a devolved government and genuine power-sharing with the constitutional nationalist party, the SDLP, there is another snag. The SDLP will not give up the link with the Republic which the Agreement has established, for this enables them to counterbalance the numerical superiority of the

Unionists. Moreover, the Republic's constitution continues to claim the entire island as the national territory. While there is no explicit support in the Republic for unification, there is enough "acquiescence in pressure for unification" among Catholics north and south of the border to keep the Irish Republican Army (IRA) in business.[16]

At the same time Britain has no interest in doing more (or less) than holding the Agreement intact. While British politicians wish to be rid of the Ulster "troubles", they cost Britain less than one per cent of its budget. To be sure, Northern Ireland is an embarrassment but there is "an acceptable level of violence", as a British politician once phrased it. The Protestant and Catholic communities have succeeded in curtailing intercommunal violence and sectarian killings. The British security forces impose limitations upon themselves in their war against the IRA. This spares lives but also ensures that the IRA retains its capacity to wreck any internal political deal between the parties representing the two groups.

The external factor has also made the conflict over Israel's occupation of the West Bank and Gaza much more intractable. For nearly two decades after the war of 1967 Israelis viewed the Palestinians in these territories as a mere fragmented extension of the Arab people whose primary foci of political and religious identification lay elsewhere. They were seen at best as an element external to Israeli society and at worst as a fifth column in Israel's struggle against the Arab states. Palestinians, in turn, tended to view Israel as a mere extension of Western imperialist power which had to be confronted through a united Arab response.[17]

Benvenisti argues (see Chapter 6) that the international conflict between Israel and the Arab states has shrunk over the last decade to a largely intercommunal conflict between Jewish Israelis and Palestinians. Nevertheless the old perceptions still impact powerfully on the politics of the Middle East.

Firstly, the conflict cannot be resolved without also addressing the larger Arab-Israeli conflict. As Prime Minister Shamir remarked in 1989: "The conflict cannot be ended without ending the conflict with the Arab states".[18] Jordan, Egypt and Syria all have different conceptions about how the conflict should be resolved, and how it should fit into the relationship between Israel and the Arab states and into inter-Arab relations. In the absence of peace with the Arab states, an independent Palestinian state represents an existential threat to the Israelis, particularly the right wing. Ariel Sharon puts it in extreme terms: "This would constitute a death sentence, for Israel's main infrastructure would be at the mercy of Palestinian and Arab missiles, rockets and artillery, as well as terrorism and a large-scale conventional invasion."[19]

Secondly, because of the conflict's international dimension politicians and diplomats continue to address the issue through diplomatic language and in the framework of international relations. This framework was ideally suited to the Israeli-Egypt peace accords of 1979 but may not fit the task of negotiating an intercommunal conflict. As Benvenisti points out, the international system is based on a recognition of its national members as legitimate, independent players interacting with other members on an equal basis. This is missing in the Israeli-Palestinian conflict: the Israeli government still refuses to deal with Palestinians as a national entity and with the PLO as its representative. As Nusseibeh's contribution makes clear, Palestinians reject any solution which does not, from the start, recognise their right to nationhood and independent statehood.

Thirdly, there is the unique status the United States enjoys as a third party. Israel enjoys a very special strategic relationship with the US, while the Palestinians rely greatly on the US to influence Israel.[20] However, despite its 3 billion dollars in annual gifts and loans to Israel the US has limited influence over Israel. Successive US administrations and Congresses have found themselves severely constrained by a powerful American Jewish lobby unwavering in its support for Israel. Furthermore, there are limits to the ability of any state, even a superpower, to pressure another state to do its bidding.

As in the case of Ulster, one has the situation where the main parties address the external players rather than their opponents in the communal conflict. Israel's obsession is to keep the support of the US and American Jewry; that of the Palestinians is to win over the Americans rather than the Israelis. As recently as January 1989 Yasser Arafat's principal associate declared: "Israel is not a principal side in the conflict for it is merely a corporation turned into a state, as has been said about General Motors . . . We are aiming at the Americans more than at the Israelis."[21] Winning international support deflects Israelis and Palestinians from building structures suitable for inter-communal bargaining.

A symmetrical exclusion process, identical to the one in Israel between Zionists and Palestinians, has been at work in South Africa. Radical opponents of the South African government have long believed that without the West investing in and trading with South Africa the regime cannot survive. On the other side, the administration of P W Botha refused to consider the ANC as anything other than an agent of Russian expansionism and communist subversion.

However, the foreign factor has never had the same impact on the conflict in South Africa. The main reason is that the adversaries lack any vital diaspora centre. Ulster Protestants and Israeli Jews have political allies and kinship ties in the Western world which are absent in the Afrikaner case. The ANC in exile tried to become the nucleus of a diaspora centre. Yet, despite the support of African states and the black lobby in the US Congress for the ANC's struggle, this organisation never enjoyed anything comparable to the support Palestinians in exile gave to the PLO or the assistance which successive Republican governments rendered to the constitutional nationalists in Ulster, culminating in the Anglo-Irish Agreement.

Towards the end of the 1980s a changing international context started to facilitate negotiations. Confronted with severe economic troubles and domestic ethnic tensions, the Soviet Union firmly indicated that it favoured a peaceful settlement in South Africa and that it was up to whites and blacks to reach accommodation. Western governments, in particular the US, discovered that their leverage over South Africa was limited. Although hurt by financial sanctions and outflow of capital the South African economy appears to have adapted.[22] Early in 1990 a spokesman for the Bush administration declared that while Washington had tilted towards the ANC in the past, the ANC was now on its own and had to secure whatever its own strength could achieve at the conference table. US Assistant Secretary of State Herman Cohen has stated policy as follows:

> Our ploy used to be putting pressure on the white power structure to end apartheid. We now feel the priority to put pressure on the white power structure and the non-white majority to engage in fruitful negotiations. We will criticise the first to move away from these negotiations.[23]

President FW de Klerk's unbanning of the liberation organisations and his offer to negotiate a new constitution were strongly influenced by a reconsideration of the external factors. The government assumes that without substantial Russian backing the ANC is a containable force. For the ANC the loss of its bases in the Frontline states has imposed the necessity of giving priority to internal political mobilisation. As a legalised organisation its challenge is now not only to build structures for political action but also to form alliances with influential white as well as black interest groups. If an intercommunal accommodation is reached it will be because South Africa has no external patron which, as in the case of Israel and Northern Ireland, can act on behalf of the internal parties.

A redefinition of identities and interests

A settlement also vitally depends on the ability of the adversaries to redefine their identities and interests. Of particular importance is a possible realignment by sections of the dominant group which could open up the possibility of a fundamental policy shift. While public opinion polls cannot be used to predict political behaviour, they provide a rough indication of a group's identity. In Northern Ireland there is little sign of a major identity change which could signal receptiveness to a major policy shift. The society has become more polarised in its sectarian divide rather than shaken up by the civil strife. Ninety per cent of the Protestants reject any form of unification with the Republic because they fear a loss of their identity.[24]

In South Africa the study of Gagiano and others shows a decrease in white-black polarisation; nevertheless upwards of 90 per cent of whites demand a state which is technocratic, capitalist, oriented towards Western standards and somehow breaks up the black majority — in short, one within which white identity is secure. As far as Israel is concerned this chapter has already commented on the powerful Jewish identity as "a people apart". Upwards of 70 per cent of Israeli Jews support the "God-and-us" and "go-it-alone" constructs which a recent study devised. This includes support for the notion "Masada will not fall again" (85 per cent), "if one rises to kill you, kill them first" (90 per cent), "Israel is and will continue to be a people dwelling alone" (69 per cent), and "world criticism stems mainly from anti-Semitism" (68 per cent).[25]

Given this cohesion, what hope is there for compromises? Here one must distinguish between coveted and acceptable solutions.[26] For the Palestinians the two most coveted solutions are a Palestinian state brought about by the expulsion of all Jews (35 per cent), and a bi-national state without expulsion of the Jews (35 per cent); however, 64 per cent of the Palestinians would accept a bi-national state without expulsion of the Jews. For Israeli Jews the most coveted option is annexation of the territories along with the expulsion of Palestinians (30 per cent). The most acceptable solution is expulsion of the Palestinians (43 per cent).[27] This, of course, would outrage world opinion and could trigger a new Middle East War.

Although polls do not indicate public support for this position as yet, the only possibility for peace probably lies in some quid pro quo in which Israel accepts the principle of Palestinian independence in exchange for Palestinian endorsement of Israel's security pre-conditions, something suggested in broad outline in Nusseibeh's chapter.

In Northern Ireland the first choice of the Protestants is integration with Britain. However, when it comes to second-best options there is a small (roughly a quarter of the community) but growing lobby for a form of devolution which entails accommodation between Protestant and Catholic representatives. Among Catholics 36 per cent favour this option. This might suggest that a devolved, power-sharing system is a viable alternative. However, as we have already seen, the Anglo-Irish Agreement has encouraged the constitutional nationalist party, the SDLP, to hold out for more. In addition, the extremist nationalist organisation, the IRA/Sinn Fein, is powerful enough to undermine or out-flank the SDLP should it move in this direction.

It is only South Africa which at this stage suggests the possibility of consensual solution. A 1986 poll conducted among all groups in the highly politicised Pretoria-Witwatersrand-Vereeniging area revealed that 75 per cent of blacks and 58 per cent of whites would accept a joint government without any group dominating.[28]

The question is why there is this political shift in South Africa but not in the other societies under discussion. The answer seems to lie in an investigation of changing identity and material concerns. In Israel, despite the *intifada*, the opposition of Jews to a Palestinian state remains strong. The great majority of Israeli Jews continue to see the territories as an external problem which should be dealt with militarily. They regard Palestinian aspirations as a threat but continue to believe strongly in Israel's ability to overcome this threat.[29] The material costs of suppressing the uprising are minimal and are outweighed by the economic benefits to Israel of a continued occupation.

In Northern Ireland the dominant Protestant community has not suffered materially as a result of the conflict. The loss of local revenue has been compensated for by the sharp rise in the British subvention — from 74m British pounds in 1968, when the troubles started, to 1,671m in 1985. The conflict has destroyed or prevented the creation of 46 000 manufacturing jobs between 1970 and 1985 but in the same period 36 000 jobs were created in the public sector (police, prisons, health, education, and so on), plus about 5 000 security-related jobs. While Protestants have suffered as a result of the loss of jobs they have fared far better than Catholics in obtaining the new jobs which opened up, particularly in the security services. The Protestant male unemployment rate of 15 per cent in the mid-1980s, as opposed to 35 per cent for Catholics, is now at a level considered normal for Catholics at the height of the economic prosperity two decades ago.[30]

Protestants have few immediate concerns about the current security situation or unfavourable long-term demographic trends. Britain does not intend to withdraw from Northern Ireland and risk creating the impression of surrender to IRA terrorism. Although the demographic gap is narrowing, Catholics will not become a majority of the population before 2026.[31] The shift of political initiative to London and Dublin has, if anything, immobilised Catholic and Protestant thinking and planning for the future.

In South Africa the situation is quite different. The rate of population increase is higher than the average growth rate of the stagnant economy. As a result the state is steadily losing its ability to provide services to the population and to curtail instability. Furthermore, whites are a shrinking percentage of the total population, down from 20 per cent in 1960 at the apogee of apartheid, to a projected 11 per cent by 2010 (Afrikaners will be down to 6 per cent). Finally, the government has faced over the past fifteen years a much more formidable mass resistance movement than its Israeli counterpart or the British security forces in Northern Ireland.

Some crucial shifts lie behind De Klerk's initiatives to normalise politics. The great majority of Afrikaners have shifted their allegiance from the *volk*, or the National Party as the guarantor of the *volk*, to the state.[32] Through the apartheid variant of indirect rule (the homelands and other apartheid bureaucracies), the state increasingly has become multiracial. As Gagiano's survey shows, white students strongly support this state. The move to bring the ANC into the system rests on the assumption that the liberation organisation will share power in government but will leave the existing state intact, or, to put it negatively, that it will not use the state to enforce the populist claims of the majority.

There is another powerful consideration, absent in the case of Northern Ireland, namely that whites have a much better chance of striking a bargain now than if they waited until their position has deteriorated further. De Klerk put this succinctly shortly after his speech of 2 February 1990:

> We did not wait until the position of power dominance turned against us before we decided to negotiate a peaceful settlement. The initiative is in our hands. We have the means to ensure that the process develops peacefully and in an orderly way.[33]

For the ANC the choice is whether to enter the system now while the economy is still strong, or to prolong the struggle and risk inheriting an irreparably damaged state and economy.

To conclude, the position of the state is crucial if any reassessment of interests and identity is to occur in a communal conflict. In Northern Ireland the absence of a state creates a vacuum in which reassessments tend to become academic exercises. In Israel the Jewish state has not only armed itself to the teeth but has invested itself with enormous symbolic power. What Israel fears is not a Palestinian state with parity in military strength but one which would attract zealots and fanatics.

There is another factor which works strongly against a policy shift. As Liebman has argued, Jewish Israelis have a very strong identification with the community. By contrast, their sense of the state, in the Western sense of the word, is weak. The state is not seen as an abstract entity with an interest of its own.[34] Hence it is extremely difficult for political elites to make unpopular strategic decisions "for reasons of state". In South Africa the position is quite different. The identification of whites with the white community has weakened while identification with the state has grown. The political elites could begin to reformulate white identity and interests because the state has steadily assumed an instrumental and utilitarian character. To a greater extent than in Israel, the white leadership can present important shifts in policy as being in the interest of the state and economy — and hence indirectly in the interests of the whites as the economically dominant community.

Over the past decade the South African state has been increasingly influenced by a multitude of think-tanks, seminars and conferences which analysed political and economic trends and options. Towards the end of the 1980s blacks began to participate in growing numbers in these discussions. No similar development occurred in Israel or Northern Ireland.

The capacity to absorb compromises

In divided societies, political institutions tend to be overburdened by the task of coping with an intercommunal conflict. However, it is usually only when leaders seek compromises that political institutions tend to fragment or collapse under the pressure. Yet without stable institutions compromises cannot be rooted within society at large.

As O'Malley explains (see Chapter 10), the Anglo-Irish Agreement was prompted to an important degree by the desire to prop up the SDLP after it had lost support. Despite this, the SDLP remains too fragile a pillar on which to base a communal compromise. In South Africa, common oppression under apartheid has long dampened intra-black tensions. However, as the race for power in a post-apartheid society accelerates, serious clashes have occurred in Natal between Inkatha, as a Zulu-based movement operating legally, and the ANC, whose leadership is predominantly Xhosa and which was long banned. Open fighting, which broke out on the Witwatersrand in August 1990, suggests a tendency for the conflict to be reduced to a straightforward Xhosa-Zulu ethnic battle which could seriously jeopardise a settlement between whites and blacks.

In Northern Ireland, South Africa and Israel parties once dominant within the ruling group have come under severe pressure from the right wing. In Israel the Labour alignment has lost out to Likud politically and ideologically. In Northern Ireland the standing of the Official Unionist Party has declined sharply from its near monopoly of Protestant support in the early 1970s. In South Africa the Conservative Party is within striking distance of unseating the NP should another white election be held again.

At the same time voters have become disaffected from the politicians. Asked if politicians had helped or hindered the peace process, 9 per cent of respondents in a Northern Ireland poll said "helped", 63 per cent said "hindered". In Israel only 14 per cent of respondents approved of the right-wing government formed by Likud's Yitzak Shamir in June 1990. An Israeli analyst noted that political action was becoming severely constrained by the fact that the country's institutional structures were "becoming more and more incapable of supporting its politics".[35]

Yet leaders have little option but to move on. The Jaffee Centre for Strategic Studies at Tel Aviv writes of the "progressive deterioration in Israel's strategic standing entailed by the status quo".[36] Unionist leaders worry about the scenario painted by MacDonald that "in staving off small concessions Protestants have hastened the day Britain will dump them on to the Republic".[37] De Klerk sees a favourable balance of power slipping away.

The dilemma for leaders is that they are up against the very historic principles and primordial fears which they or their predecessors once articulated. Ulster Unionists and Afrikaner nationalists have long proclaimed the impossibility of power-sharing with their adversaries. Nearly half of the Israeli electorate fervently believe in their right to the land and that this justifies continued occupation of the West Bank.

These beliefs are reinforced by eschatological fears, many of them irrational. For Jews there are powerful memories of the Holocaust that caution against any agreement which may remotely threaten Israel's survival. Ulster Protestants fear discrimination or, even worse, being forced off the land Catholics once occupied. A study of Afrikaner farmers in the Transvaal border regions found that upwards of eighty per cent agreed with the statement that under black rule the physical safety of whites would be threatened and that white women would be molested.[38] Adam makes a comment about South Africa that can be applied to all three societies: "It is only when the real and imagined risks of the alternative society are gradually demolished in the perception of (whites) that their greater susceptibility for and non-resistance to change can be expected."[39]

Yet compromises and political institutions that can form the bridge to an alternative society are constantly assaulted by two forces; people of violence and ideological purists. In Northern Ireland the IRA's campaign is carried out by at most 250 activists. However, as long as the IRA is in the business of violence and as long as its associate, Sinn Fein, attracts a third of the Catholic vote, the SDLP is compelled to hold out for that link with the Republic which the Protestants cannot accept. In a similar way Protestant para-militaries pressurise the constitutional parties into extreme positions. At present South Africa and Israel are free from right-wing terrorist organisations but the possibility cannot be discounted. As an Israeli analyst warns: "The combination of messianic belief and a situation of endemic national conflict has within it a built-in propensity for incremental violence — extra-legalism, vigilant-ism, selective terrorism and finally indiscriminate violence."[40]

There are also the ideological purists particularly prevalent in resistance organisations or among academics close to them, who insist that the adversaries must deny their own tradition and become absorbed in the alternative society. O'Malley phrases it well in his discussion of the attitudes of Irish Catholics: "Their willingness to allow Protestants their Protestantism does not really diminish the obligation you have to get them to see the errors of their ways; the willingness to allow Unionists their Britishness does not really diminish the obligation you have to convince them that they are really Irish."[41]

The continuing tussle between ideological commitment on the one hand and political pragmatism and compromise on the other is well reflected in the four chapters on South Africa, with each representing a different side of the coin. Schlemmer urges pragmatism for he is concerned that white leaders will be pushed into a compromise which will be wrecked by their followers who fear a betrayal of ethnic or communal claims. Adam feels that any compromise with the old order will be rejected by blacks, while Alexander fears that a compromise entrenching the capitalist system will perpetuate present racial and ethnic inequalities.

It is important to analyse the fundamental reason behind these different concerns. Both Alexander and Adam propose something akin to a final solution. Alexander accepts that his goal of a worker-led state and nation may be achieved only through revolutionary upheaval. However, he encourages alliances between democrats of all kinds to work for maximum self-organisation of the people. This may (or may not) prevent revolutionary violence. Adam, on the other hand, imagines that his solution of an ANC-led non-racial democracy will avoid turmoil by virtue of its moral superiority to the corruption of "white nationalism".

Schlemmer, by contrast, writes in the spirit of Donald Rothchild, who made this telling comment in a study of state and ethnicity in Africa:

> In an abstract sense, the termination of inter-ethnic conflict is not really a problem at all. As Adolf Hitler demonstrated so dramatically, final solutions end domestic ethnic conflicts grimly but decisively. Hence our challenge is not the elimination of conflict irrespective of cost considerations, but the minimisation of destructive encounters . . .[42]

Schlemmer senses the destructive forces lurking below the political surface in South Africa, ready to erupt were a new regime suddenly to be imposed against the will of a powerful section of the population. For him the only viable political option is a compromise which all the major parties can accept. It is out of such a compromise, rather than an ideologically purist approach, that greater political coherence can grow across the political divide.

It is also important to realise that while conditions in South Africa at the moment favour old antagonists reaching out to each other, the act of reaching a compromise is not the most difficult part: living with it is. This means finding, on an ongoing basis, a balance between national and sub-national loyalties, securing acceptance of the political rules of the game and working towards an equitable redistribution of life chances. If anything, the task is similar to that of making an unwanted marriage work. Each party will have to face up unflinchingly to its own past but compromises could be wrecked if one of the parties continued to press its case by insisting on its morally superior position while denigrating the other side's historic commitments and values. In contrast to Adam, Crick observes that there is wisdom in assuming an equal justice to the broad case put by both sides — "if one knows something of their history and grants their basic premises of argument".

Ultimately compromises and institutions can survive only if former adversaries join forces to support them and to promote a stable order that can assuage both real and irrational fears. In their work on successful transitions from authoritarian rule in Latin America and Southern Europe, O'Donnell and Schmitter develop the idea of pacts which instal regimes on a gradual basis rather than by way of dramatic event. The essence of the project is a compromise resting on the mutual acceptance that no single group is sufficiently dominant to impose its "ideal project".[43] Such pacts have led to a process in which there is gradual but steady democratisation of the budget, army and the bureaucracy.[44]

Provided it is remembered that South Africa, Israel and Northern Ireland are ethnically divided societies, unlike those discussed by O'Donnell and Schmitter, these ideas are valuable. This is particularly true of South Africa for, as Johnson suggests in his chapter, a settlement will have to be generated internally with little assistance from without.

It is on the basis of such pacts that the old ideas of nation-state and national self-determination can be reformulated. For peace can come about only if national sovereignty is shared, if nationhood is defined in pluralistic terms and if self-determination is the collective decision of the representatives of major groups, separately and together.[45]

A final word

This concluding chapter has analysed the conflicts in the three societies as they presently manifest themselves. But international and local forces impinging upon the struggle can change. Today the relative absence of external patrons favours the prospects for peace in South Africa. However, were the international community to come down strongly on one side in the negotiations, the search for peace could be severely jeopardised. The present halfway involvement of Britain and the Republic of Ireland in Ulster blocks any constructive political initiatives. but it could be that the road to peace in Northern Ireland may involve Britain withdrawing completely (much will depend on how this is done) or, at the other extreme, integrating the region fully into the United Kingdom, with the Republic formally intervening on behalf of the Ulster Catholics.

New ideological currents or the disappearance of old ones may strongly influence the internal politics of the three societies. The collapse of old-style socialism, the end of the Cold War, the vision of a united Europe and the resurgence of democracy throughout the world may push divided peoples towards seeking consensual solutions. It has contributed to the shift in the politics of South Africa and may also break down walls in Ulster and soften the divisions in the Jewish-Palestinian conflict.

Finally, institutions which are overloaded or in decay may renew themselves under inspired leadership. A striking example is Begin's use of his Likud Party to legitimise the peace settlement with Egypt. But bold and imaginative initiatives by leaders, however valuable in themselves, cannot bring peace. The essential question is whether followers can be convinced that these initiatives are in their interests and do not jeopardise their identities. In the long run this cannot be done by force or deceit. It must be done by persuasion.[46] Of all the obstacles to peace outlined in this chapter the most formidable one is the difficulty of persuading political parties and followers to move from confrontation to accommodation. If leaders fail in this endeavour, their societies will be plunged in greater turmoil than ever and hopes for vigorous economic growth will be dashed.

This chapter has argued that the weight of evidence about divided societies argues against expectations of prompt or neat solutions, and against assumptions that democratic outcomes are blocked merely by the perversity of politicians. In all three societies discussed in this book, the historic roots of the conflict are deep and the identity fears, whether real or imagined, very potent. Time and again, leaders who proposed solutions failed because the political centre was too narrow and the will to compromise too weak. Nevertheless, increasing numbers want their people not merely to survive but to survive in justice. The challenge of the 1990s is for leaders to make progress by convincing their adversaries of their desire for negotiation and peace.

ENDNOTES

1. Adrian Guelke: "Constitutional Compromises in Divided Societies", in Jesmond Blumenfeld (ed): *South Africa in Crisis*, London, Croom Helm, 1987, p151.
2. In the case of Northern Ireland the definition was in Anglo-Saxon and/or Protestant terms.
3. A distinction should be made with respect to Israeli treatment of the Arabs within the Green Line and those in the occupied territories. The former suffer some discrimination but have the vote; the latter live under colonial domination.
4. Meron Benvenisti: *Conflicts and Contradictions*, New York, Villard Books, 1986, p20.
5. John Hewitt cited by Garrett Fitzgerald: *Irish Identities*, London, BBC, 1982, p7.
6. In the case of Northern Ireland, the feeling is more of English superiority *vis à vis* "Paddy".
7. Frank Wright: *Northern Ireland: A Comparative Analysis*, Dublin, Gill and MacMillan, 1987, pp157-63.
8. Cited by W Pomeroy: "What is the National Question in International Perspective?", M van Diepen (ed): *The National Question in South Africa*, London, Zed

Books, 1988, p12. It should be observed that under the 1985-1990 states of emergency the ANC's internal associates, the UDF and MDM, tended to have more of a civil rights approach than the ANC.

9. Hamish Dickie-Clark: "The Study of Conflict in South Africa and Northern Ireland", *Social Dynamics*, Vol 2, No 1, 1976, p57.

10. Arthur Aughey: *Under Siege: Ulster Unionism and the Anglo-Irish Agreement*, Belfast, The Blackstaff Press, 1989, pp1-29.

11. Asher Arian: "A People Apart: Coping with National Security Problems in Israel", *Journal of Conflict Resolution*, Vol 33, No 4, 1989, pp603-30.

12. See also Hermann Giliomee: "The Communal Nature of the Struggle", in Hermann Giliomee and Lawrence Schlemmer (eds): *Negotiating South Africa's Future*, Johannesburg, Southern Books, 1989, pp114-29.

13. Lawrence Schlemmer: "South Africa's National Party Government", in Peter Berger and Bobby Godsell (eds): *A Future South Africa*, Cape Town, Human and Rousseau, Tafelberg, 1988, p27.

14. Wright: *Northern Ireland: A Comparative Analysis*, p275.

15. John Whyte: "Why is the Northern Ireland problem so intractable?", *Parliamentary Affairs*, XXXIV, 4, 1981, pp422-35.

16. Bob Rowthorn and Naomi Wayne: *Northern Ireland: The Political Economy of Conflict*, Cambridge, Polity Press, 1988, p135; Conor Cruise O'Brien: "A Tale of Two Nations", *The New York Review of Books*, 19 July 1990, p36. The quoted phrase is O'Brien's.

17. Meron Benvenisti: *The Shepherd's War*, Jerusalem, West Bank Data Project, 1989, pp103-5.

18. Cited by Gideon Rafael: "Middle East Talks", *International Herald Tribune*, 15 June 1989.

19. Ariel Sharon: "Making peace at a price Israel can afford", *International Herald Tribune*, 5 March 1990.

20. The Jaffee Centre for Strategic Studies, Tel Aviv University: "Israel, the West Bank and Gaza: Toward a Solution", typescript, 1989, p28.

21. Benvenisti: *The Shepherd's War*, p220.

22. The effects of the sanctions imposed since 1985 have been ambiguous. Trade sanctions have been a failure; South Africa has increased its exports in real terms more than 100 per cent. Financial sanctions have been very effective in dampening growth but there is no indication that foreign bankers will return simply because of negotiations or because a black government succeeds a white one. What both local and foreign investors demand is stability and the prospect of relative profitability. The De Klerk government hopes a NP-ANC coalition can provide this.

23. Richard Dowden: "West Offers South African Antagonists the Carrot", *The Independent*, 31 January 1990.

24. Robin Wilson: "Moving On", *Fortnight*, Belfast, November 1987, p21. Edward Moxon-Browne: *Nation, Class and Creed in Northern Ireland*, Aldershot, Gower, p126-7.

25. Arian: "National Security in Israel", p614.

26. See Michael Inbar and Ephraim Yuchtman-Yaar: "The People's Image of Conflict Resolution: Israelis and Palestinians", *Journal of Conflict Resolution*, Vol 33, No 1, 1989, p42.

27. Inbar and Yuchtman-Yaar: "The People's Image of Conflict Resolution", p46.

28. DJ van Vuuren *et al* (eds): *The South African Election of 1987*, Pinetown, Owen Burgess, 1987, p227; Hermann Giliomee and Lawrence Schlemmer: *From Apartheid to Nation-building*, Cape Town, Oxford University Press, 1990, p223.

29. Arian: "National Security in Israel", p621.

30. Rowthorn and Wayne: *Northern Ireland*, pp106-15; Kevin Boyle and Tom Hadden: *Ireland: A Positive Proposal*, Penguin Books, 1985, p3.

31. Rowthorn and Wayne: *Northern Ireland*, pp205-12.
32. For an extended discussion see Hermann Giliomee: "Afrikaner Politics, 1977–1987", John Brewer (ed): *Can South Africa Survive?*, London, MacMillan, 1989, pp108-35.
33. *Die Burger*, 31 March 1990, p2.
34. Charles Liebman: "Conceptions of 'State of Israel' in Israeli Society", *The Jerusalem Quarterly*, 47, Summer 1988, pp96-107.
35. Naomi Chazan: "Domestic Developments in Israel" in William B Quandt (ed): *The Middle East: Ten Years after Camp David*, Washington, Brookings Institution, 1988, p174; *Sunday Times*, 10 June 1990. For Northern Ireland, see Wilson: "Moving On".
36. The Jaffee Centre for Strategic Studies, Tel Aviv University: "Israel, the West Bank and Gaza: Toward a Solution", unpublished report, 1989, p28.
37. Michael MacDonald: *Children of Wrath: Political Violence in Northern Ireland*, Cambridge, Polity Press, 1986, p157.
38. Pierre Hugo: "Towards Darkness and Death: Racial Demonology in South Africa", in Pierre Hugo (ed): *South African Perspectives: Essays in Honour of Nic Olivier*, Cape Town, Die Suid-Afrikaan, 1989, pp251-5.
39. Heribert Adam: *Modernising Racial Domination: The Dynamics of South African Politics*, Berkeley, University of California Press, 1971, p118.
40. Ehud Springzak quoted by Samuel W Lewis: "Israel, Political Reality and the Search for Middle East Peace", *SAIS Review*, Winter-Spring 1987, p73.
41. For a brilliant up-to-date assessment of the Northern Ireland conflict see Padraig O'Malley: "Northern Ireland: Questions of Nuance", paper issued by the McCormack Institute of Public Affairs, University of Massachusetts, Boston. This will appear as a new concluding section in his book *The Uncivil Wars: Ireland Today*.
42. Donald Rothchild: "State and Ethnicity in Africa: A Policy Perspective", in Neil Neville and Charles Kennedy (eds): *Ethnic Preference and Public Policy in Developing States*, Boulder, Lynne Rienner Publisher, 1986, p51.
43. Guillermo O'Donnell and Philippe Schmitter: "Negotiating and Renegotiating Pacts", in Guillermo O'Donnell, *et al* (eds): *Transitions from Democratic Rule: Prospects for Democracy*, Baltimore, John Hopkins Press, 1986, pp37-40.
44. For an application to South Africa of the ideas of O'Donnell and Schmitter see F Van Zyl Slabbert: "From Domination to Democracy", *Leadership*, Vol 9, May 1990, pp66-78.
45. This draws on Bernard Crick: "Northern Ireland and the concept of consent", in Carol Harlow (ed): *Public Law and Politics*, London, Sweet and Maxwell, 1986.
46. Crick: "Northern Ireland and the concept of consent", p51.

Index